Comp.

HOME CANNING, PRESERVING, AND FREEZING

**BY
UNITED STATES DEPARTMENT OF AGRICULTURE**

Martino Publishing
Mansfield Centre, CT
2012

Martino Publishing
P.O. Box 373,
Mansfield Centre, CT 06250 USA

www.martinopublishing.com

ISBN 978-1-61427-348-6

© 2012 Martino Publishing

All rights reserved. No new contribution to this publication may
be reproduced, stored in a retrieval system, or transmitted, in any form or
by any means, electronic, mechanical, photocopying, recording, or otherwise,
without the prior permission of the Publisher.

Cover design by T. Matarazzo

Printed in the United States of America On 100% Acid-Free Paper

Complete Guide to
HOME CANNING, PRESERVING,
AND FREEZING

BY
UNITED STATES DEPARTMENT OF AGRICULTURE

United States Department of Agriculture

This edition contains unabridged reprints of the following United States Department of Agriculture Home and Garden Bulletins: No. 8, *Home Canning of Fruits and Vegetables,* revised, February 1965, slightly revised, May 1976; No. 106, *Home Canning of Meat and Poultry,* February 1966, slightly revised, Oct., 1972; No. 56, *How to Make Jellies, Jams and Preserves at Home,* revised, December 1975; No. 92, *Making Pickles and Relishes at Home,* March 1964, slightly revised, March 1970; No. 10, *Home Freezing of Fruits and Vegetables,* revised, May 1965, slightly revised, November 1971; No. 70, *Home Freezing of Poultry,* revised, January 1967, slightly revised, February 1970; No. 93, *Freezing Meat and Fish in the Home,* December 1963, slightly revised, May 1970. All of the bulletins were prepared by the Consumer and Food Economics Research Division of the Agriculture Research Service and originally published by the U.S. Government Printing Office, Washington, D.C. No. 93 was prepared in conjunction with the Animal Husbandry Research Division of the Agriculture Research Service and the Bureau of Commercial Fisheries, Fish and Wildlife Service, U.S. Department of the Interior.

CONVERSION TABLES FOR FOREIGN EQUIVALENTS

DRY INGREDIENTS

Ounces	Grams	Grams	Ounces	Pounds	Kilograms	Kilograms	Pounds
1 =	28.35	1 =	0.035	1 =	0.454	1 =	2.205
2	56.70	2	0.07	2	0.91	2	4.41
3	85.05	3	0.11	3	1.36	3	6.61
4	113.40	4	0.14	4	1.81	4	8.82
5	141.75	5	0.18	5	2.27	5	11.02
6	170.10	6	0.21	6	2.72	6	13.23
7	198.45	7	0.25	7	3.18	7	15.43
8	226.80	8	0.28	8	3.63	8	17.64
9	255.15	9	0.32	9	4.08	9	19.84
10	283.50	10	0.35	10	4.54	10	22.05
11	311.85	11	0.39	11	4.99	11	24.26
12	340.20	12	0.42	12	5.44	12	26.46
13	368.55	13	0.46	13	5.90	13	28.67
14	396.90	14	0.49	14	6.35	14	30.87
15	425.25	15	0.53	15	6.81	15	33.08
16	453.60	16	0.57				

LIQUID INGREDIENTS

Liquid Ounces	Milliliters	Milliliters	Liquid Ounces	Quarts	Liters	Liters	Quarts
1 =	29.573	1 =	0.034	1 =	0.946	1 =	1.057
2	59.15	2	0.07	2	1.89	2	2.11
3	88.72	3	0.10	3	2.84	3	3.17
4	118.30	4	0.14	4	3.79	4	4.23
5	147.87	5	0.17	5	4.73	5	5.28
6	177.44	6	0.20	6	5.68	6	6.34
7	207.02	7	0.24	7	6.62	7	7.40
8	236.59	8	0.27	8	7.57	8	8.45
9	266.16	9	0.30	9	8.52	9	9.51
10	295.73	10	0.33	10	9.47	10	10.57

Gallons (American)	Liters	Liters	Gallons (American)
1 =	3.785	1 =	0.264
2	7.57	2	0.53
3	11.36	3	0.79
4	15.14	4	1.06
5	18.93	5	1.32
6	22.71	6	1.59
7	26.50	7	1.85
8	30.28	8	2.11
9	34.07	9	2.38
10	37.86	10	2.74

CONTENTS

Home Canning of Fruits and Vegetables 1
Home Canning of Meat and Poultry 33
How to Make Jellies, Jams, and Preserves at Home . . . 57
Making Pickles and Relishes at Home 91
Home Freezing of Fruits and Vegetables 121
Home Freezing of Poultry 169
Freezing Meat and Fish in the Home 193

HOME CANNING OF FRUITS AND VEGETABLES

CONTENTS

	Page
Right canner for each food	3
Getting your equipment ready	3
General canning procedure	5
How to can fruits, tomatoes, pickled vegetables	9
Directions for fruits, tomatoes, pickled vegetables	11
How to can vegetables	16
Directions for vegetables	18
Questions and answers	29
Index	31

Acknowledgment is made to the Massachusetts Agricultural Experiment Station and the Texas Agricultural Experiment Station for cooperation in the development of some of the home-canning processes included in this publication, and to the National Canners Association for consultation and advice on processing.

Home CANNING of Fruits and Vegetables

Organisms that cause food spoilage—molds, yeasts, and bacteria—are always present in the air, water, and soil. Enzymes that may cause undesirable changes in flavor, color, and texture are present in raw fruits and vegetables.

When you can fruits and vegetables you heat them hot enough and long enough to destroy spoilage organisms. This heating (or processing) also stops the action of enzymes. Processing is done in either a boiling-water-bath canner or a steam-pressure canner. The kind of canner that should be used depends on the kind of food being canned.

Right Canner for Each Food

For fruits, tomatoes, and pickled vegetables, use a boiling-water-bath canner. You can process these acid foods safely in boiling water.

For all common vegetables except tomatoes, use a steam-pressure canner. To process these low-acid foods safely in a reasonable length of time takes a temperature higher than that of boiling water.

A pressure saucepan equipped with an accurate indicator or gage for controlling pressure at 10 pounds (240° F.) may be used as a steam-pressure canner for vegetables in pint jars or No. 2 tin cans. If you use a pressure saucepan, add 20 minutes to the processing times given in this publication for each vegetable.

Getting Your Equipment Ready

Steam-Pressure Canner

For safe operation of your canner, clean petcock and safety-valve openings by drawing a string or narrow strip of cloth through them. Do this at beginning of canning season and often during the season.

Check pressure gage.—An accurate pressure gage is necessary to get the processing temperatures needed to make food keep.

A weighted gage needs to be thoroughly clean.

A dial gage, old or new, should be checked before the canning season, and also during the season if you use the canner often. Ask your county home demonstration agent, dealer, or manufacturer about checking it.

If your gage is off 5 pounds or more, you'd better get a new one. But if the gage is not more than 4 pounds off, you can correct for it as shown below. As a reminder, tie on the canner a tag stating the reading to use to get the correct pressure.

The food is to be processed at 10 pounds steam pressure; so—

If the gage reads high—
1 pound high—process at 11 pounds.
2 pounds high—process at 12 pounds.
3 pounds high—process at 13 pounds.
4 pounds high—process at 14 pounds.

If the gage reads low—
1 pound low—process at 9 pounds.
2 pounds low—process at 8 pounds.
3 pounds low—process at 7 pounds.
4 pounds low—process at 6 pounds.

Have canner thoroughly clean.—Wash canner kettle well if you have not used it for some time. Don't put cover in water—wipe it with a soapy cloth, then with a damp, clean cloth. Dry well.

Water-Bath Canner

Water-bath canners are available on the market. Any big metal container may be used as a boiling-water-bath canner if it is deep enough so that the water is well over tops of jars and has space to boil freely. Allow 2 to 4 inches above jar tops for brisk boiling (see sketch). The canner must have a tight-fitting cover and a wire or wooden rack. If the rack has dividers, jars will not touch each other or fall against the sides of the canner during processing.

If a steam-pressure canner is deep enough, you can use it for a water bath. Cover, but do not fasten. Leave petcock wide open, so that steam escapes and pressure does not build up inside the canner.

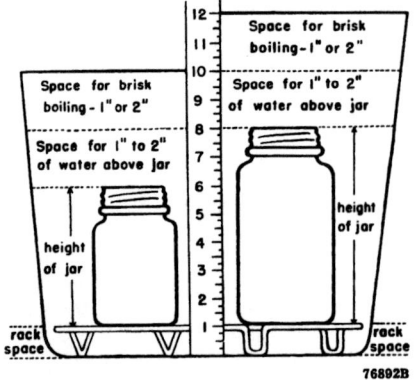

Glass Jars

Be sure all jars and closures are perfect. Discard any with cracks, chips, dents, or rust; defects prevent airtight seals.

Select the size of closure—widemouth or regular—that fits your jars.

Wash glass jars in hot, soapy water and rinse well. Wash and rinse all lids and bands. Metal lids with sealing compound may need boiling or holding in boiling water for a few minutes—follow the manufacturer's directions.

If you use rubber rings, have clean, new rings of the right size for the jars. Don't test by stretching. Wash rings in hot, soapy water. Rinse well.

Tin Cans

Select desired type and size.—Three types of tin cans are used in home canning—plain tin, C-enamel (corn enamel), and R-enamel (sanitary or standard enamel). For most products plain tin cans are satisfactory. Enameled cans are recommended for certain fruits and vegetables to prevent discoloration of food, but they are not necessary for a wholesome product.

The types of cans and the foods for which they are recommended are:

Type of can	Recommended for—
C-enamel	Corn, hominy.
R-enamel	Beets, red berries, red or black cherries, plums, pumpkin, rhubarb, winter squash.
Plain	All other fruits and vegetables for which canning directions are given in this bulletin.

In this bulletin, directions are given for canning most fruits and vegetables in No. 2 and No. 2½ tin cans. A No. 2 can holds about 2½ cups, and a No. 2½ can about 3½ cups.

Use only cans in good condition.—See that cans, lids, and gaskets are perfect. Discard badly bent, dented, or rusted cans, and lids with damaged gaskets. Keep lids in paper packing until ready to use. The paper protects the lids from dirt and moisture.

Wash cans.—Just before use, wash cans in clean water; drain upside down. Do not wash lids; washing may damage the gaskets. If lids are dusty or dirty, rinse with clean water or wipe with a damp cloth just before you put them on the cans.

Check the sealer.—Make sure the sealer you use is properly adjusted. To test, put a little water into a can, seal it, then submerge can in boiling water for a few seconds. If air bubbles rise from around the can, the seam is not tight. Adjust sealer, following manufacturer's directions.

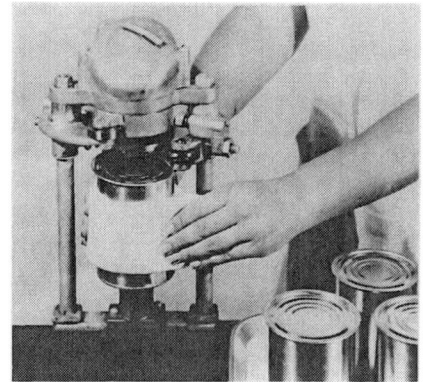

A can sealer is needed if tin cans are used.

General Canning Procedure

Selecting Fruits and Vegetables for Canning

Choose fresh, firm fruits and young, tender vegetables. Can them before they lose their freshness. If you must hold them, keep them in a cool, airy place. If you buy fruits and vegetables to can, try to get them from a nearby garden or orchard.

For best quality in the canned product, use only perfect fruits and vegetables. Sort them for size and ripeness; they cook more evenly that way.

Washing

Wash all fruits and vegetables thoroughly, whether or not they are to be pared. Dirt contains some of the bacteria hardest to kill. Wash small lots at a time, under running water or through several changes of water. Lift the food out of the water each time so dirt that has been washed off won't go back on the food. Rinse pan thoroughly between washings. Don't let fruits or vegetables soak; they may lose flavor and food value. Handle them gently to avoid bruising.

Filling Containers

Raw pack or hot pack.—Fruits and vegetables may be packed raw into glass jars or tin cans or preheated and packed hot. In this publication directions for both raw and hot packs are given for most of the foods.

Most raw fruits and vegetables should be packed tightly into the container because they shrink during processing; a few—like corn, lima beans, and peas—should be packed loosely because they expand.

Hot food should be packed fairly loosely. It should be at or near boiling temperature when it is packed.

There should be enough sirup, water, or juice to fill in around the solid food in the container and to cover the food. Food at the top of the container tends to darken if not covered with liquid. It takes from ½ to 1½ cups of liquid for a quart glass jar or a No. 2½ tin can.

Head space.—With only a few exceptions, some space should be left between the packed food and the closure. The amount of space to allow at the top of the jar or can is given in the detailed directions for canning each food.

Closing Glass Jars

Closures for glass jars are of two main types:

Metal screwband and flat metal lid with sealing compound. To use this type, wipe jar rim clean after produce is packed. Put lid on, with sealing compound next to glass. Screw metal band down tight by hand. When band is tight, this lid has enough give to let air escape during processing. Do not tighten screw band further after taking jar from canner.

Screw bands that are in good condition may be reused. You may remove bands as soon as jars are cool. Metal lids with sealing compound may be used only once.

Porcelain-lined zinc cap with shoulder rubber ring. Fit wet rubber ring down on jar shoulder, but don't stretch unnecessarily. Fill jar; wipe rubber ring and jar rim clean. Then screw cap down firmly and turn it back ¼ inch. As soon as you take jar from canner, screw cap down tight, to complete seal.

Porcelain-lined zinc caps may be reused as long as they are in good condition. Rubber rings should not be reused.

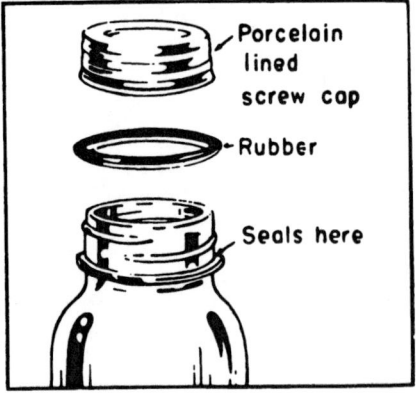

Exhausting and Sealing Tin Cans

Tin cans are sealed before processing. The temperature of the food in the cans must be 170° F. or higher when the cans are sealed. Food is heated to this temperature to drive out air so that there will be a good vacuum in the can after processing and cooling. Removal of air also helps prevent discoloring of canned food and change in flavor.

Food packed raw must be heated in the cans (exhausted) before the cans are sealed. Food packed hot may be sealed without further heating if you are sure the temperature of the food has not dropped below 170° F. To make sure, test with a thermometer, placing the bulb at the center of the can. If the thermometer registers lower than 170°, or if you do not make this test, exhaust the cans.

To exhaust, place open, filled cans on a rack in a kettle in which there is enough boiling water to come to about 2 inches below the tops of the cans. Cover the kettle. Bring water back to boiling. Boil until a thermometer inserted at the center of the can registers 170° F.—or for the length of time given in the directions for the fruit or vegetable you are canning.

Remove cans from the water one at a time, and add boiling packing liquid or water if necessary to bring head space back to the level specified for each product. Place clean lid on filled can. Seal at once.

Processing

Process fruits, tomatoes, and pickled vegetables in a boiling-water-bath canner according to the directions on page 10. Process vegetables in a steam-pressure canner according to the directions on page 16.

Cooling Canned Food

Glass jars.—As you take jars from the canner, complete seals at once if necessary. If liquid boiled out in processing, do not open jar to add more. Seal the jar just as it is.

Cool jars top side up. Give each jar enough room to let air get at all sides. Never set a hot jar on a cold surface; instead set the jars on a rack or on a folded cloth. Keep hot jars away from drafts, but don't slow cooling by covering them.

BN21476
Cool jars top side up on a rack, leaving space between jars so air can circulate.

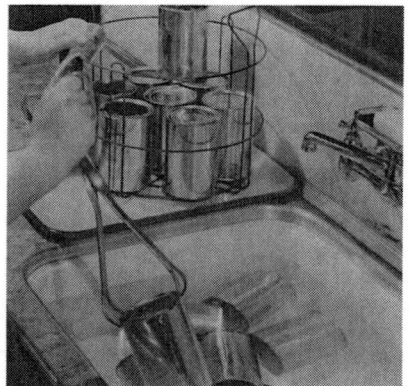
76619B
Cool tin cans in cold water; change water frequently to cool cans quickly.

Tin cans.—Put tin cans in cold, clean water to cool them; change water as needed to cool cans quickly. Take cans out of the water while they are still warm so they will dry in the air. If you stack cans, stagger them so that air can get around them.

Day-After-Canning Jobs

Test the seal on glass jars with porcelain-lined caps by turning each jar partly over in your hands. To test a jar that has a flat metal lid, press center of lid; if lid is down and will not move, jar is sealed. Or tap the center of the lid with a spoon. A clear, ringing sound means a good seal. A dull note does not always mean a poor seal; store jars without leaks and check for spoilage before use.

If you find a leaky jar, use unspoiled food right away. Or can it again; empty the jar, and pack and process food as if it were fresh. Before using jar or lid again check for defects.

When jars are thoroughly cool, take off the screw bands carefully. If a band sticks, covering for a moment with a hot, damp cloth may help loosen it.

Before storing canned food, wipe containers clean. Label to show contents, date, and lot number—if you canned more than one lot in a day.

Wash bands; store them in a dry place.

Label jars after they have been cooled.

Storing Canned Food

Properly canned food stored in a cool, dry place will retain good eating quality for a year. Canned food stored in a warm place near hot pipes, a range, or a furnace, or in direct sunlight may lose some of its eating quality in a few weeks or months, depending on the temperature.

Dampness may corrode cans or metal lids and cause leakage so the food will spoil.

Freezing does not cause food spoilage unless the seal is damaged or the jar is broken. However, frozen canned food may be less palatable than properly stored canned food. In an unheated storage place it is well to protect canned food by wrapping the jars in paper or covering them with a blanket.

On Guard Against Spoilage

Don't use canned food that shows any sign of spoilage. Look closely at each container before opening it. Bulging can ends, jar lids, or rings, or a leak—these may mean the seal has broken and the food has spoiled. When you open a container look for other signs—spurting liquid, an off odor, or mold.

It's possible for canned vegetables to contain the poison causing botulism—a serious food poisoning—without showing signs of spoilage. To avoid any risk of botulism, it is essential that the pressure canner be in perfect order and that every canning recommendation be followed exactly. Unless you're absolutely sure of your gage and canning methods, boil home-canned vegetables before tasting. Heating usually makes any odor of spoilage more evident.

Bring vegetables to a rolling boil; then cover and boil for at least 10 minutes. Boil spinach and corn 20 minutes. If the food looks spoiled, foams, or has an off odor during heating, destroy it.

Burn spoiled vegetables, or dispose of the food so that it will not be eaten by humans or animals.

How To Can Fruits, Tomatoes, Pickled Vegetables

Fruits, tomatoes, and pickled vegetables are canned according to the general directions on pages 5 to 8, the detailed directions for each food on pages 11 to 16, and the special directions given below that apply only to acid foods.

Points on Packing

Raw pack.—Put cold, raw fruits into container and cover with boiling-hot sirup, juice, or water. Press tomatoes down in the containers so they are covered with their own juice; add no liquid.

Hot pack.—Heat fruits in sirup, in water or steam, or in extracted juice before packing. Juicy fruits and tomatoes may be preheated without added liquid and packed in the juice that cooks out.

BN21474

To hot pack fruit, pack heated fruit loosely into jars.

BN21469

Cover fruit with boiling liquid before closing jar and processing in boiling-water bath.

Sweetening Fruit

Sugar helps canned fruit hold its shape, color, and flavor. Directions for canning most fruits call for sweetening to be added in the form of sugar sirup. For very juicy fruit packed hot, use sugar without added liquid.

To make sugar sirup.—Mix sugar with water or with juice extracted from some of the fruit. Use a thin, medium, or heavy sirup to suit the sweetness of the fruit and your taste. To make sirup, combine—

4 cups of water or juice.....	2 cups sugar.....	For 5 cups THIN sirup.
	3 cups sugar.....	For 5½ cups MEDIUM sirup.
	4¾ cups sugar...	For 6½ cups HEAVY sirup.

Heat sugar and water or juice together until sugar is dissolved. Skim if necessary.

To extract juice.—Crush thoroughly ripe, sound juicy fruit. Heat to simmering (185° to 210° F.) over low heat. Strain through jelly bag or other cloth.

To add sugar direct to fruit.—For juicy fruit to be packed hot, add about ½ cup sugar to each quart of raw, prepared fruit. Heat to simmering (185° to 210° F.) over low heat. Pack fruit in the juice that cooks out.

To add sweetening other than sugar.—You can use light corn sirup or mild-flavored honey to replace as much as half the sugar called for in canning fruit. Do not use brown sugar, or molasses, sorghum, or other strong-flavored sirups; their flavor overpowers the fruit flavor and they may darken the fruit.

Canning Unsweetened Fruit

You may can fruit without sweetening—in its own juice, in extracted juice, or in water. Sugar is not needed to prevent spoilage; processing is the same for unsweetened fruit as for sweetened.

Processing in Boiling-Water Bath

Directions.—Put filled glass jars or tin cans into canner containing hot or boiling water. For raw pack in glass jars have water in canner hot but not boiling; for all other packs have water boiling.

Add boiling water if needed to bring water an inch or two over tops of containers; don't pour boiling water directly on glass jars. Put cover on canner.

After jars are covered with boiling water, place lid on water-bath canner and bring water quickly back to boiling.

10 (10) Home Canning of Fruits and Vegetables

When water in canner comes to a rolling boil, start to count processing time. Boil gently and steadily for time recommended for the food you are canning. Add boiling water during processing if needed to keep containers covered.

Remove containers from the canner immediately when processing time is up.

Processing times.—Follow times carefully. The times given apply only when a specific food is prepared according to detailed directions.

If you live at an altitude of 1,000 feet or more, you have to add to these processing times in canning directions, as follows:

Altitude	Increase in processing time if the time called for is—	
	20 minutes or less	More than 20 minutes
1,000 feet	1 minute	2 minutes.
2,000 feet	2 minutes	4 minutes.
3,000 feet	3 minutes	6 minutes.
4,000 feet	4 minutes	8 minutes.
5,000 feet	5 minutes	10 minutes.
6,000 feet	6 minutes	12 minutes.
7,000 feet	7 minutes	14 minutes.
8,000 feet	8 minutes	16 minutes.
9,000 feet	9 minutes	18 minutes.
10,000 feet	10 minutes	20 minutes.

To Figure Yield of Canned Fruit From Fresh

The number of quarts of canned food you can get from a given quantity of fresh fruit depends upon the quality, variety, maturity, and size of the fruit, whether it is whole, in halves, or in slices, and whether it is packed raw or hot.

Generally, the following amounts of fresh fruit or tomatoes (as purchased or picked) make 1 quart of canned food:

	Pounds
Apples	2½ to 3
Berries, except strawberries	1½ to 3 (1 to 2 quart boxes)
Cherries (canned unpitted)	2 to 2½
Peaches	2 to 3
Pears	2 to 3
Plums	1½ to 2½
Tomatoes	2½ to 3½

In 1 pound there are about 3 medium apples and pears; 4 medium peaches or tomatoes; 8 medium plums.

Directions for Fruits, Tomatoes, Pickled Vegetables

Apples

Pare and core apples; cut in pieces. To keep fruit from darkening, drop pieces into water containing 2 tablespoons each of salt and vinegar per gallon. Drain, then boil 5 minutes in thin sirup or water.

In glass jars.—Pack hot fruit to ½ inch of top. Cover with hot sirup or water, leaving ½-inch space at top of jar. Adjust jar lids. Process in boiling-water bath (212° F.)—

Pint jars	15 minutes
Quart jars	20 minutes

As soon as you remove jars from canner, complete seals if necessary.

In tin cans.—Pack hot fruit to ¼ inch of top. Fill to top with hot sirup or water. Exhaust to 170° F. (about 10 minutes) and seal cans. Process in boiling-water bath (212° F.)—

No. 2 cans	10 minutes
No. 2½ cans	10 minutes

Applesauce

Make applesauce, sweetened or unsweetened. Heat to simmering (185°–210° F.); stir to keep it from sticking.

In glass jars.—Pack hot applesauce to ¼ inch of top. Adjust lids. Process in boiling-water bath (212° F.)—

    Pint jars_____ 10 minutes
    Quart jars_____ 10 minutes

As soon as you remove jars from canner, complete seals if necessary.

In tin cans.—Pack hot applesauce to top. Exhaust to 170° F. (about 10 minutes) and seal cans. Process in boiling-water bath (212° F.)—

    No. 2 cans_____ 10 minutes
    No. 2½ cans_____ 10 minutes

Apricots

Follow method for peaches. Peeling may be omitted.

Beets, Pickled

Cut off beet tops, leaving 1 inch of stem. Also leave root. Wash beets, cover with boiling water, and cook until tender. Remove skins and slice beets. For pickling sirup, use 2 cups vinegar (or 1½ cups vinegar and ½ cup water) to 2 cups sugar. Heat to boiling.

Pack beets in glass jars to ½ inch of top. Add ½ teaspoon salt to pints, 1 teaspoon to quarts. Cover with boiling sirup, leaving ½-inch space at top of jar. Adjust jar lids. Process in boiling-water bath (212° F.)—

    Pint jars_____ 30 minutes
    Quart jars_____ 30 minutes

As soon as you remove jars from canner, complete seals if necessary.

Berries, Except Strawberries

● **Raw Pack.**—Wash berries; drain.

In glass jars.—Fill jars to ½ inch of top. For a full pack, shake berries down while filling jars. Cover with boiling sirup, leaving ½-inch space at top. Adjust lids. Process in boiling-water bath (212° F.)—

    Pint jars_____ 10 minutes
    Quart jars_____ 15 minutes

As soon as you remove jars from canner, complete seals if necessary.

In tin cans.—Fill cans to ¼ inch of top. For a full pack, shake berries down while filling cans. Fill to top with boiling sirup. Exhaust to 170° F. (10 minutes); seal cans. Process in boiling-water bath (212° F.)—

    No. 2 cans_____ 15 minutes
    No. 2½ cans_____ 20 minutes

● **Hot Pack.**—(For firm berries)—Wash berries and drain well. Add ½ cup sugar to each quart fruit. Cover pan and bring to boil; shake pan to keep berries from sticking.

In glass jars.—Pack hot berries to ½ inch of top. Adjust jar lids. Process in boiling-water bath (212° F.)—

    Pint jars_____ 10 minutes
    Quart jars_____ 15 minutes

As soon as you remove jars from canner, complete seals if necessary.

In tin cans.—Pack hot berries to top. Exhaust to 170° F. (about 10 minutes) and seal cans. Process in boiling-water bath (212° F.)—

    No. 2 cans_____ 15 minutes
    No. 2½ cans_____ 20 minutes

Cherries

● **Raw Pack.**—Wash cherries; remove pits, if desired.

In glass jars.—Fill jars to ½ inch of top. For a full pack, shake cherries down while filling jars. Cover with boiling sirup, leaving ½-inch space at top. Adjust lids. Process in boiling-water bath (212° F.)—

    Pint jars_____ 20 minutes
    Quart jars_____ 25 minutes

As soon as you remove jars from canner, complete seals if necessary.

In tin cans.—Fill cans to ¼ inch of top. For a full pack, shake cherries down while filling cans. Fill to top with boiling sirup. Exhaust to 170° F. (about 10 minutes) and seal cans. Process in boiling-water bath (212° F.)—

No. 2 cans	20 minutes
No. 2½ cans	25 minutes

● **Hot Pack.**—Wash cherries; remove pits, if desired. Add ½ cup sugar to each quart of fruit. Add a little water to unpitted cherries to keep them from sticking while heating. Cover pan and bring to a boil.

In glass jars.—Pack hot to ½ inch of top. Adjust jar lids. Process in boiling-water bath (212° F.)—

Pint jars	10 minutes
Quart jars	15 minutes

As soon as you remove jars from canner, complete seals if necessary.

In tin cans.—Pack hot to top of cans. Exhaust to 170° F. (about 10 minutes) and seal cans. Process in boiling-water bath (212° F.)—

No. 2 cans	15 minutes
No. 2½ cans	20 minutes

Fruit Juices

Wash; remove pits, if desired, and crush fruit. Heat to simmering (185°–210° F.). Strain through cloth bag. Add sugar, if desired—about 1 cup to 1 gallon juice. Reheat to simmering.

In glass jars.—Fill jars to ½ inch of top with hot juice. Adjust lids. Process in boiling-water bath (212° F.)—

Pint jars	5 minutes
Quart jars	5 minutes

As soon as you remove jars from canner, complete seals if necessary.

In tin cans.—Fill cans to top with hot juice. Seal at once. Process in boiling-water bath (212° F.)—

No. 2 cans	5 minutes
No. 2½ cans	5 minutes

Fruit Purees

Use sound, ripe fruit. Wash; remove pits, if desired. Cut large fruit in pieces. Simmer until soft; add a little water if needed to keep fruit from sticking. Put through a strainer or food mill. Add sugar to taste. Heat again to simmering (185°–210° F.).

In glass jars.—Pack hot to ½ inch of top. Adjust lids. Process in boiling-water bath (212° F.)—

Pint jars	10 minutes
Quart jars	10 minutes

As soon as you remove jars from canner, complete seals if necessary.

In tin cans.—Pack hot to top. Exhaust to 170° F. (about 10 minutes), and seal cans. Process in boiling-water bath (212° F.)—

No. 2 cans	10 minutes
No. 2½ cans	10 minutes

Peaches

Wash peaches and remove skins. Dipping the fruit in boiling water, then quickly in cold water makes peeling easier. Cut peaches in halves; remove pits. Slice if desired. To prevent fruit from darkening during preparation, drop it into water containing 2 tablespoons each of salt and vinegar per gallon. Drain just before heating or packing raw.

BN21470

Peaches can be peeled easily if they are dipped in boiling water, then in cold water.

● **Raw Pack.**—Prepare peaches as directed above.

(Continued on next page)

Home Canning of Fruits and Vegetables (13)

In glass jars.—Pack raw fruit to ½ inch of top. Cover with boiling sirup, leaving ½-inch space at top of jar. Adjust jar lids. Process in boiling-water bath (212° F.)—

>   Pint jars_____ 25 minutes
>   Quart jars_____ 30 minutes

As soon as you remove jars from canner, complete seals if necessary.
In tin cans.—Pack raw fruit to ¼ inch of top. Fill to top with boiling sirup. Exhaust to 170° F. (about 10 minutes) and seal cans. Process in boiling-water bath (212° F.)—

>   No. 2 cans_____ 30 minutes
>   No. 2 ½ cans_____ 35 minutes

● **Hot Pack.**—Prepare peaches as directed above. Heat peaches through in hot sirup. If fruit is very juicy you may heat it with sugar, adding no liquid.
In glass jars.—Pack hot fruit to ½ inch of top. Cover with boiling liquid, leaving ½-inch space at top of jar. Adjust jar lids. Process in boiling-water bath (212° F.)—

>   Pint jars_____ 20 minutes
>   Quart jars_____ 25 minutes

As soon as you remove jars from canner, complete seals if necessary.
In tin cans.—Pack hot fruit to ¼ inch of top. Fill to top with boiling liquid. Exhaust to 170° F. (about 10 minutes) and seal cans. Process in boiling-water bath (212° F.)—

>   No. 2 cans_____ 25 minutes
>   No. 2½ cans_____ 30 minutes

Pears

Wash pears. Peel, cut in halves, and core. Continue as with peaches, either raw pack or hot pack.

Plums

Wash plums. To can whole, prick skins. Freestone varieties may be halved and pitted.
● **Raw Pack.**—Prepare plums as directed above.
In glass jars.—Pack raw fruit to ½ inch of top. Cover with boiling sirup, leaving ½-inch space at top of jar. Adjust jar lids. Process in boiling-water bath (212° F.)—

>   Pint jars_____ 20 minutes
>   Quart jars_____ 25 minutes

As soon as you remove jars from canner, complete seals if necessary.
In tin cans.—Pack raw fruit to ¼ inch of top. Fill to top with boiling sirup. Exhaust to 170° F. (about 10 minutes) and seal cans. Process in boiling-water bath (212° F.)—

>   No. 2 cans_____ 15 minutes
>   No. 2½ cans_____ 20 minutes

● **Hot Pack.**—Prepare plums as directed above. Heat to boiling in sirup or juice. If fruit is very juicy you may heat it with sugar, adding no liquid.
In glass jars.—Pack hot fruit to ½ inch of top. Cover with boiling liquid, leaving ½-inch space at top of jar. Adjust jar lids. Process in boiling-water bath (212° F.)—

>   Pint jars_____ 20 minutes
>   Quart jars_____ 25 minutes

As soon as you remove jars from canner, complete seals if necessary.
In tin cans.—Pack hot fruit to ¼ inch of top. Fill to top with boiling liquid. Exhaust to 170° F. (about 10 minutes) and seal cans. Process in boiling-water bath (212° F.)—

>   No. 2 cans_____ 15 minutes
>   No. 2½ cans_____ 20 minutes

Rhubarb

Wash rhubarb and cut into ½-inch pieces. Add ½ cup sugar to each quart rhubarb and let stand to draw out juice. Bring to boiling.
In glass jars.—Pack hot to ½ inch of top. Adjust lids. Process in boiling-water bath (212° F.)—

>   Pint jars_____ 10 minutes
>   Quart jars_____ 10 minutes

As soon as you remove jars from canner, complete seals if necessary.
In tin cans.—Pack hot to top of cans. Exhaust to 170° F. (about 10 minutes) and seal cans. Process in boiling-water bath (212° F.)—

No. 2 cans_____ 10 minutes
No. 2½ cans_____ 10 minutes

Tomatoes

Use only firm, ripe, red tomatoes. Do not use overripe tomatoes, because tomatoes lose acidity as they mature. Tomatoes with soft spots or decayed areas are not suitable for canning. To loosen skins, dip into boiling water for about ½ minute; then dip quickly into cold water. Cut out stem ends and peel tomatoes.

To peel tomatoes, dip them in boiling water, then quickly in cold water to loosen skins.

To raw pack tomatoes, put peeled tomatoes in jars and press down to fill spaces.

● **Raw Pack.**—Leave tomatoes whole or cut in halves or quarters.

In glass jars.—Pack tomatoes to ½ inch of top, pressing gently to fill spaces. Add no water. Add ½ teaspoon salt to pints; 1 teaspoon to quarts. Adjust lids. Process in boiling-water bath (212° F.)—

Pint jars_____ 35 minutes
Quart jars_____ 45 minutes

As soon as you remove jars from canner, complete seals if necessary.

In tin cans.—Pack tomatoes to top of cans, pressing gently to fill spaces. Add no water. Add ½ teaspoon salt to No. 2 cans; 1 teaspoon to No. 2½ cans. Exhaust to 170° F., (about 15 minutes) and seal cans. Process in boiling-water bath (212° F.)—

No. 2 cans_____ 45 minutes
No. 2½ cans_____ 55 minutes

● **Hot Pack.**—Quarter peeled tomatoes. Bring to boil; stir to keep tomatoes from sticking.

In glass jars.—Pack boiling-hot tomatoes to ½ inch of top. Add ½ teaspoon salt to pints; 1 teaspoon to quarts. Adjust jar lids. Process in boiling-water bath (212° F.)—

Pint jars_____ 10 minutes
Quart jars_____ 10 minutes

As soon as you remove jars from canner, complete seals if necessary.

In tin cans.—Pack boiling-hot tomatoes to ¼ inch of top. Add no water. Add ½ teaspoon salt to No. 2 cans; 1 teaspoon to No. 2½ cans. Exhaust to 170° F. (about 10 minutes) and seal cans. Process in boiling-water bath (212° F.)—

No. 2 cans_____ 10 minutes
No. 2½ cans_____ 10 minutes

Tomato Juice

Use ripe, juicy, red tomatoes. Do not use overripe tomatoes, because tomatoes lose acidity as they mature. Tomatoes with soft or decayed areas are not suitable for canning. Wash, remove stem ends, cut into pieces. Simmer until softened, stirring often. Put through strainer. Add 1 teaspoon salt to each quart juice. Reheat at once just to boiling.

Home Canning of Fruits and Vegetables (15)

In glass jars.—Fill jars with boiling-hot juice to ½ inch of top. Adjust jar lids. Process in boiling-water bath (212° F.)—

    Pint jars_____ 10 minutes
    Quart jars_____ 10 minutes

As soon as you remove jars from canner, complete seals if necessary.

In tin cans.—Fill cans to top with boiling-hot juice. Seal cans at once. Process in boiling-water bath (212° F.)—

    No. 2 cans_____ 15 minutes
    No. 2½ cans_____ 15 minutes

How To Can Vegetables

Can vegetables according to general directions on pages 5 to 8, the detailed directions for each vegetable on pages 18 to 28. and special directions below that apply only to vegetables.

Points on Packing

Raw pack.—Pack cold raw vegetables (except corn, lima beans, and peas) tightly into container and cover with boiling water.

Hot pack.—Preheat vegetables in water or steam. Cover with cooking liquid or boiling water. Cooking liquid is recommended for packing most vegetables because it may contain minerals and vitamins dissolved out of the food. Boiling water is recommended when cooking liquid is dark, gritty, or strong-flavored, and when there isn't enough cooking liquid.

Processing in a Pressure Canner

Use a steam-pressure canner for processing all vegetables except tomatoes and pickled vegetables. A pressure saucepan may be used for pint jars and No. 2 cans (see p. 3).

Directions.—Follow the manufacturer's directions for the canner you are using. Here are a few pointers on the use of any steam-pressure canner:

● Put 2 or 3 inches of boiling water in the bottom of the canner; the amount of water to use depends on the size and shape of the canner.

● Set filled glass jars or tin cans on rack in canner so that steam can flow around each container. If two layers of cans or jars are put in, stagger the second layer. Use a rack between layers of glass jars.

● Fasten canner cover securely so that no steam can escape except through vent (petcock or weighted-gage opening).

● Watch until steam pours steadily from vent. Let it escape for 10 minutes or more to drive all air from the canner. Then close petcock or put on weighted gage.

● Let pressure rise to 10 pounds (240° F.). The moment this pressure is reached start counting processing time. Keep pressure constant by regulating heat under the canner. Do not lower pressure by opening petcock. Keep drafts from blowing on canner.

● When processing time is up, remove canner from heat immediately.

With glass jars, let canner stand until pressure is zero. Never try to rush the cooling by pouring cold water over the canner. When pressure registers zero, wait a minute or two, then slowly open petcock or take off weighted gage. Unfasten cover and tilt the far side up so steam escapes away from you. Take jars from canner.

To process vegetables, bring pressure in canner up to 10 pounds, then start to count processing time.

With tin cans, release steam in canner as soon as canner is removed from heat by opening petcock or taking off weighted gage. Then take off canner cover and remove cans.

Processing times.—Follow processing times carefully. The times given apply only when a specific food is prepared according to detailed directions.

If you live at an altitude of less than 2,000 feet above sea level, process vegetables at 10 pounds pressure for the times given.

At altitudes above sea level, it takes more than 10 pounds pressure to reach 240° F. If you live at an altitude of 2,000 feet, process vegetables at 11 pounds pressure. At 4,000 feet, use 12 pounds pressure; at 6,000 feet, 13 pounds pressure; at 8,000 feet, 14 pounds pressure; at 10,000 feet, 15 pounds pressure.

A weighted gage may need to be corrected for altitude by the manufacturer.

To Figure Yield of Canned Vegetables From Fresh

The number of quarts of canned food you can get from a given amount of fresh vegetables depends on quality, condition, maturity, and variety of the vegetable, size of pieces, and on the way the vegetable is packed—raw or hot pack.

Generally, the following amounts of fresh vegetables (as purchased or picked) make 1 quart of canned food:

	Pounds		Pounds
Asparagus	2½ to 4½	Okra	1½
Beans, lima, in pods	3 to 5	Peas, green, in pods	3 to 6
Beans, snap	1½ to 2½	Pumpkin or winter squash	1½ to 3
Beets, without tops	2 to 3½	Spinach and other greens	2 to 6
Carrots, without tops	2 to 3	Squash, summer	2 to 4
Corn, sweet, in husks	3 to 6	Sweetpotatoes	2 to 3

Directions for Vegetables

Asparagus

● **Raw Pack.**—Wash asparagus; trim off scales and tough ends and wash again. Cut into 1-inch pieces.

In glass jars.—Pack asparagus as tightly as possible without crushing to ½ inch of top. Add ½ teaspoon salt to pints; 1 teaspoon to quarts. Cover with boiling water, leaving ½-inch space at top of jar. Adjust jar lids. Process in pressure canner at 10 pounds pressure (240° F.)—

Pint jars	25 minutes
Quart jars	30 minutes

As soon as you remove jars from canner, complete seals if necessary.

In tin cans.—Pack asparagus as tightly as possible without crushing to ¼ inch of top. Add ½ teaspoon salt to No. 2 cans; 1 teaspoon to No. 2½ cans. Fill to top with boiling water. Exhaust to 170° F. (about 10 minutes) and seal cans. Process in pressure canner at 10 pounds pressure (240° F.)—

No. 2 cans	20 minutes
No. 2½ cans	20 minutes

● **Hot Pack.**—Wash asparagus; trim off scales and tough ends and wash again. Cut in 1-inch pieces; cover with boiling water. Boil 2 or 3 minutes.

In glass jars.—Pack hot asparagus loosely to ½ inch of top. Add ½ teaspoon salt to pints; 1 teaspoon to quarts. Cover with boiling-hot cooking liquid, or if liquid contains grit use boiling water. Leave ½-inch space at top of jar. Adjust jar lids. Process in pressure canner at 10 pounds pressure (240° F.)—

Pint jars	25 minutes
Quart jars	30 minutes

As soon as you remove jars from canner, complete seals if necessary.

In tin cans.—Pack hot asparagus loosely to ¼ inch of top. Add ½ teaspoon salt to No. 2 cans; 1 teaspoon to No. 2½ cans. Fill to top with boiling-hot cooking liquid, or if liquid contains grit use boiling water. Exhaust to 170° F. (about 10 minutes) and seal cans. Process in pressure canner at 10 pounds pressure (240° F.)—

No. 2 cans	20 minutes
No. 2½ cans	20 minutes

Beans, Dry, With Tomato or Molasses Sauce

Sort and wash dry beans (kidney, navy, or yellow eye). Cover with boiling water; boil 2 minutes, remove from heat and let soak 1 hour. Heat to boiling, drain, and save liquid for making sauce.

In glass jars.—Fill jars three-fourths full with hot beans. Add a small piece of salt pork, ham, or bacon. Fill to 1 inch of top with hot sauce (see recipes below). Adjust jar lids. Process in pressure canner at 10 pounds pressure (240° F.)—

Pint jars	65 minutes
Quart jars	75 minutes

As soon as you remove jars from canner, complete seals if necessary.

In tin cans.—Fill cans three-fourths full with hot beans. Add a small piece of salt pork, ham, or bacon. Fill to ¼ inch of top with hot sauce (see recipes below). Exhaust to 170° F. (about 20 minutes) and seal cans. Process in pressure canner at 10 pounds pressure (240° F.)—

No. 2 cans	65 minutes
No. 2½ cans	75 minutes

Tomato sauce.—Mix 1 quart tomato juice, 3 tablespoons sugar, 2 teaspoons salt, 1 tablespoon chopped onion, and ¼ teaspoon mixture of ground cloves, allspice, mace, and cayenne. Heat to boiling.

Or mix 1 cup tomato catsup with 3 cups of water or soaking liquid from beans and heat to boiling.

Molasses sauce.—Mix 1 quart water or soaking liquid from beans, 3 tablespoons dark molasses, 1 tablespoon vinegar, 2 teaspoons salt, and ¾ teaspoon powdered dry mustard. Heat to boiling.

Beans, Dry, Baked

Soak and boil beans according to directions for beans with sauce.

Place small pieces of salt pork, ham, or bacon in earthenware crock or a pan.

Add beans. Add enough molasses sauce to cover beans. Cover crock and bake 4 to 5 hours at 350° F. (moderate oven). Add water as needed—about every hour.

In glass jars.—Pack hot beans to 1 inch of top. Adjust jar lids. Process in pressure canner at 10 pounds pressure (240° F.)—

Pint jars	80 minutes
Quart jars	100 minutes

As soon as you remove jars from canner, complete seals if necessary.

In tin cans.—Pack hot beans to ¼ inch of top. Exhaust to 170° F. (about 15 minutes) and seal cans. Process in pressure canner at 10 pounds pressure (240° F.)—

No. 2 cans	95 minutes
No. 2½ cans	115 minutes

Beans, Fresh Lima

Can only young, tender beans.

● **Raw Pack.**—Shell and wash beans.

In glass jars.—Pack raw beans into clean jars. For small-type beans, fill to 1 inch of top of jar for pints and 1½ inches for quarts; for large beans, fill to ¾ inch of top for pints and 1¼ inches for quarts. Beans should not be pressed or shaken down. Add ½ teaspoon salt to pints; 1 teaspoon to quarts. Fill jar to ½ inch of top with boiling water. Adjust jar lids. Process in pressure canner at 10 pounds pressure (240° F.)—

Pint jars	40 minutes
Quart jars	50 minutes

As soon as you remove jars from canner, complete seals if necessary.

In tin cans.—Pack raw beans to ¾ inch of top; do not shake or press beans down. Add ½ teaspoon salt to No. 2 cans; 1 teaspoon to No. 2½ cans. Fill cans to top with boiling water. Exhaust to 170° F. (about 10 minutes) and seal cans. Process in pressure canner at 10 pounds pressure (240° F.)—

No. 2 cans	40 minutes
No. 2½ cans	40 minutes

● **Hot Pack.**—Shell the beans, cover with boiling water, and bring to boil.

In glass jars.—Pack hot beans loosely to 1 inch of top. Add ½ teaspoon salt to pints; 1 teaspoon to quarts. Cover with boiling water, leaving 1-inch space at top of jar. Adjust jar lids. Process in pressure canner at 10 pounds pressure (240° F.)—

Pint jars	40 minutes
Quart jars	50 minutes

As soon as you remove jars from canner, complete seals if necessary.

In tin cans.—Pack hot beans loosely to ½ inch of top. Add ½ teaspoon salt to No. 2 cans; 1 teaspoon to No. 2½ cans. Fill to top with boiling water. Exhaust to 170° F. (about 10 minutes) and seal cans. Process in pressure canner at 10 pounds pressure (240° F.)—

No. 2 cans	40 minutes
No. 2½ cans	40 minutes

Beans, Snap

● **Raw Pack.**—Wash beans. Trim ends; cut into 1-inch pieces.

In glass jars.—Pack raw beans tightly to ½ inch of top. Add ½ teaspoon salt to pints; 1 teaspoon to quarts. Cover with boiling water, leaving ½-inch space at top of jar. Adjust jar lids. Process in pressure canner at 10 pounds pressure (240° F.)—

Pint jars	20 minutes
Quart jars	25 minutes

As soon as you remove jars from canner, complete seals if necessary.

In tin cans.—Pack raw beans tightly to ¼ inch of top. Add ½ tea-

(Continued on next page)

spoon salt to No. 2 cans; 1 teaspoon to No. 2½ cans. Fill to top with boiling water. Exhaust to 170° F. (about 10 minutes) and seal cans. Process in pressure canner at 10 pounds pressure (240° F.)—

    No. 2 cans_____ 25 minutes
    No. 2½ cans_____ 30 minutes

● **Hot Pack.**—Wash beans. Trim ends; cut into 1-inch pieces. Cover with boiling water; boil 5 minutes.

In glass jars.—Pack hot beans loosely to ½ inch of top. Add ½ teaspoon salt to pints; 1 teaspoon to quarts. Cover with boiling-hot cooking liquid, leaving ½-inch space at top of jar. Adjust jar lids. Process in pressure canner at 10 pounds pressure (240° F.)—

    Pint jars_____ 20 minutes
    Quart jars_____ 25 minutes

As soon as you remove jars from canner, complete seals if necessary.

In tin cans.—Pack hot beans loosely to ¼ inch of top. Add ½ teaspoon salt to No. 2 cans; 1 teaspoon to No. 2½ cans. Fill to top with boiling-hot cooking liquid. Exhaust to 170° F. (about 10 minutes) and seal cans. Process in pressure canner at 10 pounds pressure (240° F.)—

    No. 2 cans_____ 25 minutes
    No. 2½ cans_____ 30 minutes

Beets

Sort beets for size. Cut off tops, leaving an inch of stem. Also leave root. Wash beets. Cover with boiling water and boil until skins slip easily—15 to 25 minutes, depending on size. Skin and trim. Leave baby beets whole. Cut medium or large beets in ½-inch cubes or slices; halve or quarter very large slices.

In glass jars.—Pack hot beets to ½ inch of top. Add ½ teaspoon salt to pints; 1 teaspoon to quarts. Cover with boiling water, leaving ½-inch space at top of jar. Adjust jar lids. Process in pressure canner at 10 pounds pressure (240° F.)—

    Pint jars_____ 30 minutes
    Quart jars_____ 35 minutes

As soon as you remove jars from canner, complete seals if necessary.

In tin cans.—Pack hot beets to ¼ inch of top. Add ½ teaspoon salt to No. 2 cans; 1 teaspoon to No. 2½ cans. Fill to top with boiling water. Exhaust to 170° F. (about 10 minutes) and seal cans. Process in pressure canner at 10 pounds pressure (240° F.)—

    No. 2 cans_____ 30 minutes
    No. 2½ cans_____ 30 minutes

BN21475

To hot pack snap beans, cover cut beans with boiling water and boil 5 minutes.

BN21471

Then pack hot beans loosely in jar and cover with hot cooking liquid before processing in a pressure canner.

Beets, Pickled
See page 12.

Carrots

● **Raw Pack.**—Wash and scrape carrots. Slice or dice.

In glass jars.—Pack raw carrots tightly into clean jars, to 1 inch of top of jar. Add ½ teaspoon salt to pints; 1 teaspoon to quarts. Fill jar to ½ inch of top with boiling water. Adjust jar lids. Process in pressure canner at 10 pounds pressure (240° F.)—

    Pint jars_____ 25 minutes
    Quart jars_____ 30 minutes

As soon as you remove jars from canner, complete seals if necessary.

In tin cans.—Pack raw carrots tightly into cans to ½ inch of top. Add ½ teaspoon salt to No. 2 cans; 1 teaspoon to No. 2½ cans. Fill cans to top with boiling water. Exhaust to 170° F. (about 10 minutes) and seal cans. Process in pressure canner at 10 pounds pressure (240° F.)—

    No. 2 cans_____ 25 minutes
    No. 2½ cans_____ 30 minutes

● **Hot Pack.**—Wash and scrape carrots. Slice or dice. Cover with boiling water and bring to boil.

In glass jars.—Pack hot carrots to ½ inch of top. Add ½ teaspoon salt to pints; 1 teaspoon to quarts. Cover with boiling-hot cooking liquid, leaving ½-inch space at top of jar. Adjust jar lids. Process in pressure canner at 10 pounds pressure (240° F.)—

    Pint jars_____ 25 minutes
    Quart jars_____ 30 minutes

As soon as you remove jars from canner, complete seals if necessary.

In tin cans.—Pack hot carrots to ¼ inch of top. Add ½ teaspoon salt to No. 2 cans; 1 teaspoon to No. 2½ cans. Fill with boiling-hot cooking liquid. Exhaust to 170° F. (about 10 minutes) and seal cans. Process in pressure canner at 10 pounds pressure (240° F.)—

    No. 2 cans_____ 20 minutes
    No. 2½ cans_____ 25 minutes

Corn, Cream-Style

● **Raw Pack.**—Husk corn and remove silk. Wash. Cut corn from cob at about center of kernel and scrape cobs.

In glass jars.—Use pint jars only. Pack corn to 1½ inches of top; do not shake or press down. Add ½ teaspoon salt to each jar. Fill to ½ inch of top with boiling water. Adjust jar lids. Process in pressure canner at 10 pounds pressure (240° F.)—

    Pint jars_____ 95 minutes

As soon as you remove jars from canner, complete seals if necessary.

In tin cans.—Use No. 2 cans only. Pack corn to ½ inch of top; do not shake or press down. Add ½ teaspoon salt to each can. Fill cans to top with boiling water. Exhaust to 170° F. (about 25 minutes) and seal cans. Process in pressure canner at 10 pounds pressure (240° F.)—

    No. 2 cans_____ 105 minutes

● **Hot Pack.**—Husk corn and remove silk. Wash. Cut corn from cob at about center of kernel and scrape cob. To each quart of corn add 1 pint boiling water. Heat to boiling.

In glass jars.—Use pint jars only. Pack hot corn to 1 inch of top. Add ½ teaspoon salt to each jar. Adjust jar lids. Process in pressure canner at 10 pounds pressure (240° F.)—

    Pint jars_____ 85 minutes

As soon as you remove jars from canner, complete seals if necessary.

In tin cans.—Use No. 2 cans only. Pack hot corn to top. Add ½ teaspoon salt to each can. Exhaust to 170° F. (about 10 minutes) and seal cans. Process in pressure canner at 10 pounds pressure (240° F.)—

    No. 2 cans_____ 105 minutes

Corn, Whole-Kernel

● **Raw Pack.**—Husk corn and remove silk. Wash. Cut from cob at about two-thirds the depth of kernel.

In glass jars.—Pack corn to 1 inch of top; do not shake or press down.

(*Continued on next page*)

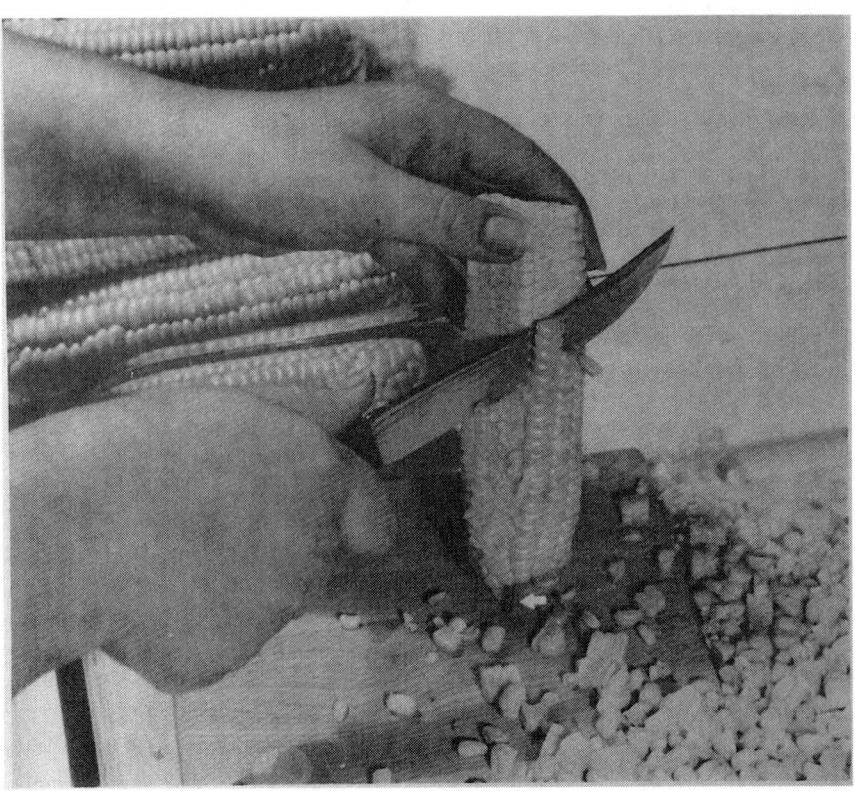

A nail driven at an angle through the cutting board (see arrow) holds the cob steady and makes it easy to cut corn from the cob.

Add ½ teaspoon salt to pints; 1 teaspoon to quarts. Fill to ½ inch of top with boiling water. Adjust jar lids. Process in pressure canner at 10 pounds pressure (240° F.)—

    Pint jars_____ 55 minutes
    Quart jars_____ 85 minutes

As soon as you remove jars from canner, complete seals if necessary.

In tin cans.—Pack corn to ½ inch of top; do not shake or press down. Add ½ teaspoon salt to No. 2 cans; 1 teaspoon to No. 2½ cans. Fill to top with boiling water. Exhaust to 170° F. (about 10 minutes) and seal cans. Process in pressure canner at 10 pounds pressure (240° F.)—

    No. 2 cans_____ 60 minutes
    No. 2½ cans_____ 60 minutes

● **Hot Pack.**—Husk corn and remove silk. Wash. Cut from cob at about two-thirds the depth of kernel. To each quart of corn add 1 pint boiling water. Heat to boiling.

In glass jars.—Pack hot corn to 1 inch of top and cover with boiling-hot cooking liquid, leaving 1-inch space at top of jar. Or fill to 1 inch of top with mixture of corn and liquid. Add ½ teaspoon salt to pints; 1 teaspoon to quarts. Adjust jar lids. Process in pressure canner at 10 pounds pressure (240° F.)—

    Pint jars_____ 55 minutes
    Quart jars_____ 85 minutes

As soon as you remove jars from canner, complete seals if necessary.

To hot pack corn, put heated corn loosely in C-enamel cans; fill cans with boiling liquid.

In tin cans.—Pack hot corn to ½ inch of top and fill to top with boiling-hot cooking liquid. Or fill to top with mixture of corn and liquid. Add ½ teaspoon salt to No. 2 cans; 1 teaspoon to No. 2½ cans. Exhaust to 170° F. (about 10 minutes) and seal cans. Process in pressure canner at 10 pounds pressure (240° F.)—

No. 2 cans_____ 60 minutes
No. 2½ cans_____ 60 minutes

Hominy

Place 2 quarts of dry field corn in an enameled pan; add 8 quarts of water and 2 ounces of lye. Boil vigorously ½ hour, then allow to stand for 20 minutes. Rinse off the lye with several hot water rinses. Follow with cold water rinses to cool for handling.

Work hominy with the hands until dark tips of kernels are removed (about 5 minutes). Separate the tips from the corn by floating them off in water or by placing the corn in a coarse sieve and washing thoroughly. Add sufficient water to cover hominy about 1 inch, and boil 5 minutes; change water. Repeat 4 times. Then cook until kernels are soft (½ to ¾ hour) and drain. This will make about 6 quarts of hominy.

In glass jars.—Pack hot hominy to ½ inch of top. Add ½ teaspoon salt to pints; 1 teaspoon to quarts. Cover with boiling water, leaving ½-inch space at top of jar. Adjust jar lids. Process in pressure canner at 10 pounds pressure (240° F.)—

Pint jars_____ 60 minutes
Quart jars_____ 70 minutes

As soon as you remove jars from canner, complete seals if necessary.

In tin cans.—Pack hot hominy to ¼ inch of top. Add ½ teaspoon salt to No. 2 cans; 1 teaspoon to No. 2½ cans. Fill to top with boiling water. Exhaust to 170° F. (about 10 minutes) and seal cans. Process in pressure canner at 10 pounds pressure (240° F.)—

No. 2 cans_____ 60 minutes
No. 2½ cans_____ 70 minutes

Mushrooms

Trim stems and discolored parts of mushrooms. Soak mushrooms in cold water for 10 minutes to remove adhering soil. Wash in clean water. Leave small mushrooms whole; cut larger ones in halves or quarters. Steam 4 minutes or heat gently for 15 minutes without added liquid in a covered saucepan.

In glass jars.—Pack hot mushrooms to ½ inch of top. Add ¼ teaspoon salt to half pints; ½ teaspoon to pints. For better color, add crystalline ascorbic acid—$\frac{1}{16}$ teaspoon to half-pints; ⅛ teaspoon to pints. Add boiling-hot cooking liquid or boiling water to cover mushrooms, leaving ½-inch space at top of jar. Adjust jar lids. Process in pressure canner at 10 pounds pressure (240° F.)—

Half-pint jars_____ 30 minutes
Pint jars_____ 30 minutes

As soon as you remove jars from canner, complete seals if necessary.

In tin cans.—Pack hot mushrooms to ¼ inch of top of cans. Add ½ teaspoon salt to No. 2 cans. For better color, add ⅛ teaspoon of crystalline ascorbic acid to No. 2 cans. Then fill to top with boiling-hot cooking liquid or boiling water. Exhaust to 170° F. (about 10 minutes) and seal cans. Process in pressure canner at 10 pounds pressure (240° F.)—

No. 2 cans_____ 30 minutes

Okra

Can only tender pods. Wash; trim. Cook for 1 minute in boiling water. Cut into 1-inch lengths or leave pods whole.

In glass jars.—Pack hot okra to ½ inch of top. Add ½ teaspoon salt to pints; 1 teaspoon to quarts. Cover with boiling water, leaving ½-inch space at top of jar. Adjust jar lids. Process in pressure canner at 10 pounds pressure (240° F.)—

    Pint jars_____ 25 minutes
    Quart jars_____ 40 minutes

As soon as you remove jars from canner, complete seals if necessary.

In tin cans.—Pack hot okra to ¼ inch of top. Add ½ teaspoon salt to No. 2 cans; 1 teaspoon to No. 2½ cans. Fill to top with boiling water. Exhaust to 170° F. (about 10 minutes) and seal cans. Process in pressure canner at 10 pounds pressure (240° F.)—

    No. 2 cans_____ 25 minutes
    No. 2½ cans_____ 35 minutes

Peas, Fresh Blackeye (Cowpeas, Blackeye Beans)

● **Raw Pack.**—Shell and wash blackeye peas.

In glass jars.—Pack raw blackeye peas to 1½ inches of top of pint jars and 2 inches of top of quart jars; do not shake or press peas down. Add ½ teaspoon salt to pints; 1 teaspoon to quarts. Cover with boiling water, leaving ½-inch space at top of jars. Adjust jar lids. Process in pressure canner at 10 pounds pressure (240° F.)—

    Pint jars_____ 35 minutes
    Quart jars_____ 40 minutes

As soon as you remove jars from canner, complete seals if necessary.

In tin cans.—Pack raw blackeye peas to ¾ inch of top; do not shake or press down. Add ½ teaspoon salt to No. 2 cans; 1 teaspoon to No. 2½ cans. Cover with boiling water, leaving ¼-inch space at top of cans. Exhaust to 170° F. (about 10 minutes) and seal cans. Process in pressure canner at 10 pounds pressure (240° F.)—

    No. 2 cans_____ 35 minutes
    No. 2½ cans_____ 40 minutes

● **Hot Pack.**—Shell and wash blackeye peas, cover with boiling water, and bring to a rolling boil. Drain.

In glass jars.—Pack hot blackeye peas to 1¼ inches of top of pint jars and 1½ inches of top of quart jars; do not shake or press peas down. Add ½ teaspoon salt to pints; 1 teaspoon to quarts. Cover with boiling water, leaving ½-inch space at top of jar. Adjust jar lids. Process in pressure canner at 10 pounds pressure (240° F.)—

    Pint jars_____ 35 minutes
    Quart jars_____ 40 minutes

As soon as you remove jars from canner, complete seals if necessary.

In tin cans.—Pack hot blackeye peas to ½ inch of top; do not shake or press peas down. Add ½ teaspoon salt to No. 2 cans; 1 teaspoon to No. 2½ cans. Cover with boiling water, leaving ¼-inch space at top of cans. Exhaust to 170° F. (about 10 minutes) and seal cans. Process in pressure canner at 10 pounds pressure (240° F.)—

    No. 2 cans_____ 30 minutes
    No. 2½ cans_____ 35 minutes

Peas, Fresh Green

● **Raw Pack.**—Shell and wash peas.

In glass jars.—Pack peas to 1 inch of top; do not shake or press down. Add ½ teaspoon salt to pints; 1 teaspoon to quarts. Cover with boiling water, leaving 1½ inches of space at top of jar. Adjust jar lids. Process in pressure canner at 10 pounds pressure (240° F.)—

    Pint jars_____ 40 minutes
    Quart jars_____ 40 minutes

As soon as you remove jars from canner, complete seals if necessary.

In tin cans.—Pack peas to ¼ inch of top; do not shake or press down. Add ½ teaspoon salt to No. 2 cans; 1 teaspoon to No. 2½ cans. Fill to top with boiling water. Exhaust to 170° F. (about 10 minutes) and seal cans. Process at 10 pounds pressure (240° F.)—

    No. 2 cans_____ 30 minutes
    No. 2½ cans_____ 35 minutes

● **Hot Pack.**—Shell and wash peas. Cover with boiling water. Bring to boil.

In glass jars.—Pack hot peas loosely to 1 inch of top. Add ½ teaspoon salt to pints; 1 teaspoon to quarts. Cover with boiling water, leaving 1-inch space at top of jar. Adjust jar lids. Process in pressure canner at 10 pounds pressure (240° F.)—

    Pint jars_____ 40 minutes
    Quart jars_____ 40 minutes

As soon as you remove jars from canner, complete seals if necessary.

In tin cans.—Pack hot peas loosely to ¼ inch of top. Add ½ teaspoon salt to No. 2 cans; 1 teaspoon to No. 2½ cans. Fill to top with boiling water. Exhaust to 170° F. (about 10 minutes) and seal cans. Process at 10 pounds pressure (240° F.)—

    No. 2 cans_____ 30 minutes
    No. 2½ cans_____ 35 minutes

Potatoes, Cubed

Wash, pare, and cut potatoes into ½-inch cubes. Dip cubes in brine (1 teaspoon salt to 1 quart water) to prevent darkening. Drain. Cook for 2 minutes in boiling water, drain.

In glass jars.—Pack hot potatoes to ½ inch of top. Add ½ teaspoon salt to pints; 1 teaspoon to quarts. Cover with boiling water, leaving ½-inch space at top of jar. Adjust jar lids. Process in pressure canner at 10 pounds pressure (240° F.)—

    Pint jars_____ 35 minutes
    Quart jars_____ 40 minutes

As soon as you remove jars from canner, complete seals if necessary.

In tin cans.—Pack hot potatoes to ¼ inch of top. Add ½ teaspoon salt to No. 2 cans; 1 teaspoon to No. 2½ cans. Fill to top with boiling water. Exhaust to 170° F. (about 10 minutes) and seal cans. Process in pressure canner at 10 pounds pressure (240° F.)—

    No. 2 cans_____ 35 minutes
    No. 2½ cans_____ 40 minutes

Potatoes, Whole

Use potatoes 1 to 2½ inches in diameter. Wash, pare, and cook in boiling water for 10 minutes. Drain.

In glass jars.—Pack hot potatoes to ½ inch of top. Add ½ teaspoon salt to pints; 1 teaspoon to quarts. Cover with boiling water, leaving ½-inch space at top of jar. Adjust jar lids. Process in pressure canner at 10 pounds pressure (240° F.)—

    Pint jars_____ 30 minutes
    Quart jars_____ 40 minutes

As soon as you remove jars from canner, complete seals if necessary.

In tin cans.—Pack hot potatoes to ¼ inch of top. Add ½ teaspoon salt to No. 2 cans; 1 teaspoon to No. 2½ cans. Fill to top with boiling water. Exhaust to 170° F. (about 10 minutes) and seal cans. Process in pressure canner at 10 pounds pressure (240° F.)—

    No. 2 cans_____ 35 minutes
    No. 2½ cans_____ 40 minutes

Pumpkin, Cubed

Wash pumpkin, remove seeds, and pare. Cut into 1-inch cubes. Add just enough water to cover; bring to boil.

In glass jars.—Pack hot cubes to ½ inch of top. Add ½ teaspoon salt to pints; 1 teaspoon to quarts. Cover with hot cooking liquid, leaving ½-inch space at top of jar. Adjust jar lids. Process in pressure canner at 10 pounds pressure (240° F.)—

    Pint jars_____ 55 minutes
    Quart jars_____ 90 minutes

(Continued on next page)

As soon as you remove jars from canner, complete seals if necessary.

In tin cans.—Pack hot cubes to ¼ inch of top. Add ½ teaspoon salt to No. 2 cans; 1 teaspoon to No. 2½ cans. Fill to top with hot cooking liquid. Exhaust to 170° F. (about 10 minutes) and seal cans. Process in pressure canner at 10 pounds pressure (240° F.)—

  No. 2 cans_____ 50 minutes
  No. 2½ cans_____ 75 minutes

Pumpkin, Strained

Wash pumpkin, remove seeds, and pare. Cut into 1-inch cubes. Steam until tender, about 25 minutes. Put through food mill or strainer. Simmer until heated through; stir to keep pumpkin from sticking to pan.

In glass jars.—Pack hot to ½ inch of top. Add no liquid or salt. Adjust jar lids. Process at 10 pounds pressure (240° F.)—

  Pint jars_____ 65 minutes
  Quart jars_____ 80 minutes

As soon as you remove jars from canner, complete seals if necessary.

In tin cans.—Pack hot to ⅛ inch of top. Add no liquid or salt. Exhaust to 170° F. (about 10 minutes) and seal cans. Process in pressure canner at 10 pounds pressure (240° F.)—

  No. 2 cans_____ 75 minutes
  No. 2½ cans_____ 90 minutes

Spinach (and Other Greens)

Can only freshly picked, tender spinach. Pick over and wash thoroughly. Cut out tough stems and midribs. Place about 2½ pounds of spinach in a cheesecloth bag and steam about 10 minutes or until well wilted.

In glass jars.—Pack hot spinach loosely to ½ inch of top. Add ¼ teaspoon salt to pints; ½ teaspoon to quarts. Cover with boiling water, leaving ½-inch space at top of jar. Adjust jar lids. Process in pressure canner at 10 pounds pressure (240° F.)—

  Pint jars_____ 70 minutes
  Quart jars_____ 90 minutes

As soon as you remove jars from canner, complete seals if necessary.

In tin cans.—Pack hot spinach loosely to ¼ inch of top. Add ¼ teaspoon salt to No. 2 cans; ½ teaspoon to No. 2½ cans. Fill to top with boiling water. Exhaust to 170° F. (about 10 minutes) and seal cans. Process in pressure canner at 10 pounds pressure (240° F.)—

  No. 2 cans_____ 65 minutes
  No. 2½ cans_____ 75 minutes

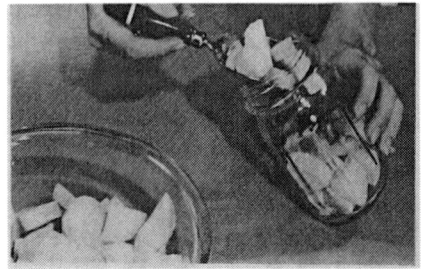

78351B

To raw pack squash, pack uniform pieces of squash tightly into jars.

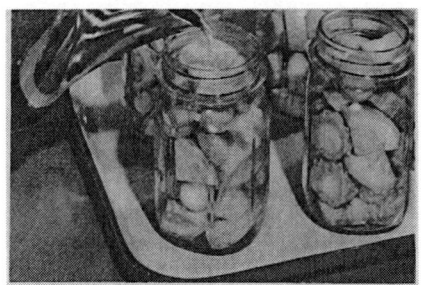

78352B

Cover squash with boiling water just before closing jars and putting in pressure canner.

Squash, Summer

● **Raw Pack.**—Wash but do not pare squash. Trim ends. Cut squash into ½-inch slices; halve or quarter to make pieces of uniform size.

In glass jars.—Pack raw squash tightly into clean jars to 1 inch of top of jar. Add ½ teaspoon salt to pints; 1 teaspoon to quarts. Fill jar to ½

When processing time is up, let pressure in canner drop to zero. Slowly open petcock or take off weighted gage. Unfasten cover, tilting far side up so steam escapes away from you.

inch of top with boiling water. Adjust jar lids. Process in pressure canner at 10 pounds pressure (240° F.)—

Pint jars_____ 25 minutes
Quart jars_____ 30 minutes

As soon as you remove jars from canner, complete seals if necessary.

In tin cans.—Pack raw squash tightly into cans to ½ inch of top. Add ½ teaspoon salt to No. 2 cans; 1 teaspoon to No. 2½ cans. Fill cans to top with boiling water. Exhaust to 170° F. (about 10 minutes) and seal cans. Process in pressure canner at 10 pounds pressure (240° F.)—

No. 2 cans_____ 20 minutes
No. 2½ cans_____ 20 minutes

● **Hot Pack.**—Wash squash and trim ends; do not pare. Cut squash into ½-inch slices; halve or quarter to make pieces of uniform size. Add just enough water to cover. Bring to boil.

In glass jars.—Pack hot squash loosely to ½ inch of top. Add ½ teaspoon salt to pints; 1 teaspoon to quarts. Cover with boiling-hot cooking liquid, leaving ½-inch space at top of jar. Adjust jar lids. Process in pressure canner at 10 pounds pressure (240° F.)—

Pint jars_____ 30 minutes
Quart jars_____ 40 minutes

As soon as you remove jars from canner, complete seals if necessary.

(*Continued on next page*)

Home Canning of Fruits and Vegetables (27)

In tin cans.—Pack hot squash loosely to ¼ inch of top. Add ½ teaspoon salt to No. 2 cans; 1 teaspoon to No. 2½ cans. Fill to top with boiling-hot cooking liquid. Exhaust to 170° F. (about 10 minutes) and seal cans. Process in pressure canner at 10 pounds pressure (240° F.)—

    No. 2 cans_____ 20 minutes
    No. 2½ cans_____ 20 minutes

Squash, Winter

Follow method for pumpkin.

Sweetpotatoes, Dry Pack

Wash sweetpotatoes. Sort for size. Boil or steam until partially soft (20 to 30 minutes). Skin. Cut in pieces if large.

In glass jars.—Pack hot sweetpotatoes tightly to 1 inch of top, pressing gently to fill spaces. Add no salt or liquid. Adjust jar lids. Process in pressure canner at 10 pounds pressure (240° F.)—

    Pint jars_____ 65 minutes
    Quart jars_____ 95 minutes

As soon as you remove jars from canner, complete seals if necessary.

In tin cans.—Pack hot sweetpotatoes tightly to top of can, pressing gently to fill spaces. Add no salt or liquid. Exhaust to 170° F. (about 10 minutes) and seal cans. Process in pressure canner at 10 pounds pressure (240° F.)—

    No. 2 cans_____ 80 minutes
    No. 2½ cans_____ 95 minutes

Sweetpotatoes, Wet Pack

Wash sweetpotatoes. Sort for size. Boil or steam just until skins slip easily. Skin and cut in pieces.

In glass jars.—Pack hot sweetpotatoes to 1 inch of top. Add ½ teaspoon salt to pints; 1 teaspoon to quarts. Cover with boiling water or medium sirup, leaving 1-inch space at top of jar. Adjust jar lids. Process in pressure canner at 10 pounds pressure (240° F.)—

    Pint jars_____ 55 minutes
    Quart jars_____ 90 minutes

As soon as you remove jars from canner, complete seals if necessary.

In tin cans.—Pack hot sweetpotatoes to ¼ inch of top. Add ½ teaspoon salt to No. 2 cans; 1 teaspoon to No. 2½ cans. Fill to top with boiling water or medium sirup. Exhaust to 170° F. (about 10 minutes) and seal cans. Process in pressure canner at 10 pounds pressure (240° F.)—

    No. 2 cans_____ 70 minutes
    No. 2½ cans_____ 90 minutes

Questions and Answers

Q. *Is it safe to process foods in the oven?*
A. No, oven canning is dangerous. Jars may explode. The temperature of the food in jars during oven processing does not get high enough to insure destruction of spoilage bacteria in vegetables.

Q. *Why is open-kettle canning not recommended for fruits and vegetables?*
A. In open-kettle canning, food is cooked in an ordinary kettle, then packed into hot jars and sealed without processing. For vegetables, the temperatures obtained in open-kettle canning are not high enough to destroy all the spoilage organisms that may be in the food. Spoilage bacteria may get in when the food is transferred from kettle to jar.

Q. *May a pressure canner be used for processing fruits and tomatoes?*
A. Yes. If it is deep enough it may be used as a water-bath canner (p. 4). Or you may use a pressure canner to process fruits and tomatoes at 0 to 1 pound pressure without having the containers of food completely covered with water. Put water in the canner to the shoulders of the jars; fasten cover. When live steam pours steadily from the open vent, start counting time. Leave vent open and process for the same times given for the boiling-water bath.

Q. *Must glass jars and lids be sterilized by boiling before canning?*
A. No, not when boiling-water bath or pressure-canner method is used. The containers as well as the food are sterilized during processing. But be sure jars and lids are clean.

Q. *Why is liquid sometimes lost from glass jars during processing?*
A. Loss of liquid may be due to packing jars too full, fluctuating pressure in a pressure canner, or lowering pressure too suddenly.

Q. *Should liquid lost during processing be replaced?*
A. No, never open a jar and refill with liquid—this would let in bacteria and you would need to process again. Loss of liquid does not cause food to spoil, though the food above the liquid may darken.

Q. *Is it safe to use home canned food if liquid is cloudy?*
A. Cloudy liquid may be a sign of spoilage. But it may be caused by the minerals in hard water, or by starch from overripe vegetables. If liquid is cloudy, boil the food. Do not taste or use any food that foams during heating or has an off odor.

Q. *Why does canned fruit sometimes float in jars?*
A. Fruit may float because pack is too loose or sirup too heavy; or because some air remains in tissues of the fruit after heating and processing.

Q. *Is it safe to can foods without salt?*
A. Yes. Salt is used for flavor only and is not necessary for safe processing.

Q. What makes canned foods change color?

A. Darkening of foods at the tops of jars may be caused by oxidation due to air in the jars or by too little heating or processing to destroy enzymes. Overprocessing may cause discoloration of foods throughout the containers.

Pink and blue colors sometimes seen in canned pears, apples, and peaches are caused by chemical changes in the coloring matter of the fruit.

Iron and copper from cooking utensils or from water in some localities may cause brown, black, and gray colors in some foods.

When canned corn turns brown, the discoloring may be due to the variety of corn, to stage of ripeness, to overprocessing, or to copper or iron pans.

Packing liquid may dissolve coloring materials from the foods. The use of plain tin cans will cause some foods to lose color (p. 4).

Q. Is it safe to eat discolored canned foods?

A. The color changes noted above do not mean the food is unsafe to eat. However, spoilage may also cause color changes. Any canned food that has an unusual color should be examined carefully before use (p. 8).

Q. Does ascorbic acid help keep fruits and vegetables from darkening?

A. Yes. The addition of ¼ teaspoon of crystalline ascorbic acid (vitamin C) to a quart of fruit or vegetable before it is processed retards oxidation, which is one cause of darkening of canned foods. One teaspoon of crystalline ascorbic acid weighs about 3 grams (or 3,000 milligrams).

Q. Is it all right to use preservatives in home canning?

A. No. Some canning powders or other chemical preservatives may be harmful.

Q. Why do the undersides of metal lids sometimes discolor?

A. Natural compounds in some foods corrode the metal and make a brown or black deposit on the underside of the lid. This deposit is harmless.

Q. When canned or frozen fruits are bought in large containers, is it possible to can them in smaller containers?

A. Any canned or frozen fruit may be heated through, packed, and processed the same length of time as recommended for hot packs. This canned food may be of lower quality than if fruit had been canned when fresh.

Q. Is it safe to leave food in tin cans after opening?

A. Yes. Food in tin cans needs only to be covered and refrigerated.

Q. Is the processing time the same no matter what kind of range is used?

A. Processing times and temperatures in this bulletin are for canning in a pressure canner or boiling-water bath with any type of range.

Q. Can fruits and vegetables be canned without heating if aspirin is used?

A. No. Aspirin cannot be relied on to prevent spoilage or to give satisfactory products. Adequate heat treatment is the only safe procedure.

INDEX

	Page
Altitude, high, canning at	11, 17
Apples	11
Applesauce	12
Apricots	12
Ascorbic acid	23, 30
Asparagus	18
Aspirin	30
Beans, dry	18, 19
Beans, fresh lima	19
Beans, snap	19
Beets	20
Beets, pickled	12
Berries	12
Botulinus poisoning	8
Carrots	21
Cherries	12
Corn	21
Discoloration, causes	30
Frozen foods, canning	30
FRUIT—	
floating, cause	29
how to can	9
juices	13
preparation for canning	5
purees	13
yield of canned from fresh	11
Gage	3
GLASS JARS—	
filling	5, 9, 16
preparing for use	4
processing	4, 17
sealing	6
sterilization	29
Greens	26
Hominy	23
Honey, used in canning fruit	10
JUICES—	
fruit	13
tomato	15
Leaks, tests for	8
Lima beans, fresh	19
LIQUID—	
cloudy	29
cooking, when to use	16
loss in canning	29
Mushrooms	23
Okra	24
Open-kettle canning	29
Oven canning	29

	Page
PACKING—	
fruits	5, 9
hot pack	5, 9, 16
raw pack	5, 9, 16
vegetables	5, 16
Peaches	13
Pears	14
Peas, fresh blackeye	24
Peas, fresh green	24
Plums	14
Potatoes	25
Preservatives	30
Pressure saucepan	3
Pumpkin	25
Purees, fruit	13
Recanning fruit	30
Rhubarb	14
Rubber rings	4, 6
Salt, canning without	29
Sealer, tin-can, testing	5
Seals, testing	8
Sirup, corn	10
Sirup, sugar	9
Spoilage	8, 30
Spinach	26
Squash	26
STEAM-PRESSURE CANNER—	
care	3
for processing fruit	29
processing in	16
Storage	8
SUGAR—	
canning without	10
in canning fruit	9
Sweetpotatoes	28
TIN CANS—	
exhausting	6
filling	5, 16
holding food in opened can	30
preparing for use	5
processing	17
types	4
Tomatoes	15
Tomato juice	15
VEGETABLES—	
how to can	16
preparation for canning	5
yield of canned from fresh	17
WATER-BATH CANNER—	
preparing for use	4
processing in	10

HOME CANNING OF
MEAT AND POULTRY

Contents

	Page		Page
Getting ready	4	Spoilage	13
Meats	4	Directions for meat	18
Equipment	4	Cut-up meat	18
Canning methods	8	Ground meat	19
Packing	9	Sausage	19
Closing jars	10	Corned beef	19
Sealing cans	10	Meat-vegetable stew	20
Processing	11	Heart and tongue	20
Yield of canned meat from fresh	11	Soup stock	20
After-canning jobs	12	Directions for poultry	21
Cooling	12	Cut-up poultry	21
Checking seals	12	Giblets	22
Labeling	12	Questions and answers	23
Storing	13	Index	24

Home Canning of Meat and Poultry

Prepared by
CONSUMER AND FOOD ECONOMICS RESEARCH DIVISION
Agricultural Research Service

Fresh, wholesome meats and fresh, wholesome poultry are suitable for home canning. Frozen meats also may be canned at home.

Popular meats for home canning are—
- Beef, veal, mutton. lamb, pork.
- Chicken, duck, goose, guinea, squab, turkey.
- Rabbit.
- Game birds.
- Small-game animals.
- Large-game animals.

Meat and poultry canned at home must be processed in a pressure canner. Either glass jars or tin cans may be used for home canning.

To insure the safety and wholesomeness of the meats you can at home—
- Start with good-quality fresh or frozen meat.
- Keep all meat, work surfaces, and equipment clean.
- Make sure the pressure canner is in good working condition.
- Pack and close containers carefully.
- Process meat for recommended time.
- Test seals after cooling containers.
- Label containers.
- Store canned meat in cool, dry place.

Acknowledgment is made to the research laboratories of the National Canners Association for consultation and advice on processing.

Follow all canning directions carefully. Processing times and temperatures were developed specifically for use with a pressure canner.

Meat may contain bacteria that cause botulism, a severe form of food poisoning. These bacteria are destroyed when cans or jars of food are processed at a temperature of 240° F. for the times specified.

There is a risk of botulism from home-canned meats if the processing temperature is lower than 240° F. or if processing time is shorter than recommended.

It is not safe to process canned meat in a boiling-water bath, an oven, a steamer without pressure, or an open kettle. None of these methods will heat the meat enough to kill dangerous bacteria in a reasonable time.

There also is a risk of botulism if shortcuts are taken in canning meats, if untested directions are used, or if processing times (pp. 18 to 23) are changed.

Getting Ready

Meats

Use only good-quality meat or poultry—home-produced or purchased from a farm or store.

Chill home-produced meat immediately after slaughter to prevent spoiling and to permit tenderizing. Meat is easier to handle when it is cold. For thorough chilling, keep meat at a temperature below 40° F. until time to prepare it for canning; can it within a few days after slaughter.

If refrigeration is not available and if the maximum daily temperature is above 40° F., process the meat as soon as body heat is gone.

If meat must be held for longer than a few days, freeze it. Store frozen meat at temperatures of 0° F. or lower until canning time. Then cut or saw frozen meat into pieces of desired size.

If frozen meat is thawed before canning, thaw it in a refrigerator at a temperature of 40° F. or lower until most of the ice crystals have disappeared.

Keep all meat clean and sanitary. Rinse poultry thoroughly in cold water, then drain.

Keep all meat as cool as possible during preparation for canning. Handle it rapidly; process it as soon as containers are packed.

Equipment

To control the bacteria that cause spoilage, keep everything that touches meat as clean as possible.

Scrub metal, enamelware, and porcelain pans in hot soapy water. Rinse pans well in boiling water before putting meat in them. Wash knives and kitchen tools to be used in canning; rinse well with boiling water.

Cutting boards, wood utensils, and wooden work surfaces need special treatment to keep spoilage bacteria under control. Scrape surfaces if necessary; scrub with hot soapy water and rinse well with boiling water. Then disinfect clean surfaces.

For disinfecting, use a liquid chlorine disinfectant (household laundry bleach) or other disinfectant. Dilute according to directions on the container. Cover wooden surfaces with the disinfectant solution and leave 15 minutes. Wash solution off with boiling water.

Pressure canner

To insure the safety of canned meats and poultry, jars or cans must be processed at a sufficiently high temperature for a long enough time to kill all bacteria that cause spoilage or food poisoning.

The only practical way to get this high temperature is to use a pressure canner. When steam is held under 10 pounds of pressure at sea level, the temperature in the canner quickly reaches 240° F.—the necessary safe temperature for canning meat.

A pressure canner should be equipped with a rack to hold jars or cans.

A pressure saucepan with accurate controls may be safely used for processing meats in pint jars or No. 2 cans. If you use a pressure saucepan, add 20 minutes to the processing times specified.

Before using the canner, wash the kettle well. Do not put cover with dial gage in water. Wipe the cover carefully with a hot soapy cloth; repeat with a clean damp cloth. Dry. Keep the petcock and safety valve clear. Before each use of the canner, inspect these openings. To clean, draw a string or narrow strip of cloth through the petcock.

Pressure adjustments.—If you live above sea level, you may need to adjust steam pressure in your canner to get a temperature of 240° F. The rule: For each 2,000 feet above sea level, increase the pressure by 1 pound.

CAUTION: Do not increase processing time when you increase steam pressure.

If a weighted gage is used at a high altitude, have it corrected for altitude by the manufacturer of the canner.

Gage adjustments.—When a weighted gage is adjusted for altitude, it needs no further regulation.

A dial gage should be checked before the canning season. If you use the canner frequently, have the gage checked several times a year. Ask your extension home economist, your dealer, or the manufacturer about checking the accuracy of a dial gage.

If the dial gage is not accurate, tie a warning tag to the canner. On the tag, write the margin of error, the date the canner was tested, and the gage setting to use for the correct pressure (see below).

All directions in this bulletin require processing at 10 pounds of steam pressure. The following adjustments give the correct pressure:

If the gage reads high—
    1 pound high—process at 11 pounds.
    2 pounds high—process at 12 pounds.
    3 pounds high—process at 13 pounds.
    4 pounds high—process at 14 pounds.

If the gage reads low—
    1 pound low—process at 9 pounds.
    2 pounds low—process at 8 pounds.
    3 pounds low—process at 7 pounds.
    4 pounds low—process at 6 pounds.

It is not safe to use a canner if the dial gage registers as much as 5 pounds high or low. Replace a faulty gage with an accurate one.

Glass jars

Jars may be widemouth or regular type. Pints and quarts are satisfactory sizes for canning meats and poultry.

There are two types of jar closures:
- Flat metal lid with sealing compound and a metal screw band. This closure seals as jar cools.
- Porcelain-lined zinc cap with a shoulder rubber ring. This cap must be tightened to complete the seal immediately after meat is canned.

Be sure all jars and closures are perfect. Discard jars with cracks or chips; discard lids and bands with dents or rust. Defects prevent air-tight seals.

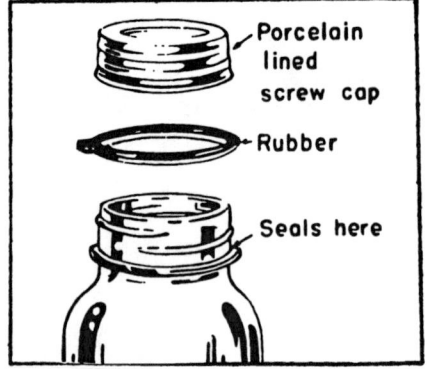

Before canning, jars must be thoroughly clean. It is not necessary to sterilize them before they are filled, however. Processing at the recommended steam pressure sterilizes both the containers and their contents.

Wash jars and lids in hot soapy water; rinse well. Some metal lids with sealing compound may need boiling or holding in boiling water for a few minutes before use. Do not reuse metal lids with sealing compounds. Follow the manufacturer's directions.

If you use rubber rings, get new ones of the right size to fit jars. Don't test by stretching. Wash rings in hot soapy water. Rinse well.

Tin cans

Use plain tin cans in good condition for canning meats.

C-enamel, R-enamel, and sanitary-enamel cans are not suitable for meat. Fat in meat or poultry may cause enamel to peel off the inside of the can. Meat in such cans appears unappetizing, but it is not harmful.

Make sure that cans, lids, and gaskets are perfect. Discard cans that are badly bent, dented, or rusty; discard lids with damaged gaskets.

Protect lids from dirt and moisture by storing them in original packing.

Directions are given for canning meat in No. 2 and No. 2½ tin cans. A No. 2 can holds 2½ cups; a No. 2½ can holds 3½ cups.

Wash cans in clean hot water just before use. Drain upside down.

Do not wash lids of cans, because washing may damage the gaskets. If lids have become soiled, rinse with clean water or wipe with a damp cloth when you are ready to put them on cans.

Sealer

If you use cans, you need a sealer in good working order. Before processing meat or poultry, adjust sealer according to manufacturer's directions.

The finished seam between lid and can should be smooth and even.

Test by sealing a can containing a small amount of water. Submerge the sealed can in boiling water for a few seconds. If air bubbles rise from around the can, the seam is not tight. Readjust the sealer.

Thermometer

It is a good idea to use a thermometer both when meat is packed hot and when the canning directions call for removing (exhausting) air from jars or cans. With a thermometer, you are able to make sure meat is heated to 170° F.—the minimum temperature needed to exhaust air properly.

Place thermometer in center of jar or can that is being heated. The thermometer bulb should be about half-way to the bottom of the container.

If a thermometer is not available, follow the times given in the directions.

For hot pack, pour boiling liquid over packed poultry or meat before closing jar and processing in a pressure canner.

Canning Methods

Prepare and process meat and poultry according to general directions given. Directions specify the types of packs and types of containers suitable for each meat product listed. Instructions must be followed carefully to assure a product safe from spoilage.

Detailed directions for canning meats are given on pages 18 to 20. Picture sequences are shown on pages 14 and 15.

Detailed directions for canning poultry are given on pages 21 to 23. Pictures illustrating main steps are on pages 16 and 17.

How To Make Broth

To make meat or poultry broth, place bony pieces in saucepan and cover with cold water. Simmer until meat is tender. Pour broth into another pan; skim off fat. Add boiling broth to containers packed with precooked meat or poultry; fill to level specified in directions.

Packing

Pack meat loosely in containers. Jars may lose liquid during processing if they are packed too tightly or too full.

Work with one glass jar or tin can at a time. Keep precooked meat hot while packing. Use boiling liquid—broth, meat juice, or water—if directions call for added liquid.

Two methods are used for packing meat:

- *Hot pack.* Meat is precooked before it is packed in jars or cans. Boiling broth or boiling water is poured over meat before containers are processed in a pressure canner. (See p. 8 for directions on how to make broth.) The temperature of food packed hot should be at least 170° F. at the time jars are closed or cans are sealed.
- *Raw pack.* Meat is packed uncooked. Raw-packed meat usually is heated to 170° F. to exhaust—or remove—air from jars or cans before processing in a pressure canner.

Directions for canning in glass jars require exhausting air from raw-packed meat products, except meat-vegetable stew and raw pack poultry, with bone. These two products may be processed without exhausting if they are raw packed in glass jars according to directions on pages 20 and 22.

Directions for using tin cans include exhausting air from all raw-packed meat. It always is necessary to exhaust air from raw-packed meat in tin cans before processing because air has no way to escape after cans are sealed.

Exhausting air

To exhaust—or remove—air, set open jars or cans packed with raw meat on a rack in a large pan of boiling water. Water level should be about 2 inches below tops of jars or cans. Cover the pan. Cook meat in containers at slow boil until temperature at center of jars or cans registers 170° F. If a thermometer is not available, follow times given to cook meat until medium done.

When raw-packed meat is heated to 170° F., air is driven out of the food so that a vacuum will be formed in jars or cans after processing and cooling. Exhausting air also helps to prevent changes in the flavor of canned meat.

Salt

Salt may be added to canned meat for flavor. It does not act as a preservative in canned meat, so it is not needed to make the product safe.

If you decide to use salt, add it after meat is packed in the jar or can. Amounts for various sized containers are given in the canning directions.

Fat

Remove as much fat as possible from meat before canning. Cut off all large lumps; trim marbled meat without slashing the lean unnecessarily.

Do not use excessively fat meat or poultry for canning.

After packing containers, wipe the tops free of fat. Any fat that gets on the rim of jars or cans may prevent an airtight seal.

Closing Jars

If jar has a flat metal lid: Wipe rim of packed jar to remove fat and meat particles that might prevent a proper seal. Put lid on jar with sealing compound next to glass. Screw the metal band down tight by hand. When band is screwed tight, this lid has enough "give" to let air escape during processing. Do not tighten band further after taking jar from canner.

If jar has a porcelain-lined zinc lid: Fit wet rubber ring down on shoulder of empty jar. Don't stretch ring unnecessarily. Pack jar with meat. Wipe rubber ring and jar rim clean. Then screw cap down firmly and turn it back ¼ inch before processing. As soon as you take jar from canner, screw cap down tight to complete seal.

Sealing Cans

Use a can sealer in good working condition. Follow the manufacturer's directions carefully.

Wipe rim clean; place lid on can. Seal at once.

A can sealer is needed if tin cans are used.

Processing

Use a pressure canner for processing meat. A pressure saucepan may be used for pint jars or No. 2 cans (see p. 5).

Follow the manufacturer's directions carefully. Here are a few suggestions about using a pressure canner:

- Put 2 or 3 inches of water in the canner; heat to boiling. Use enough water to prevent the canner from boiling dry.
- Set packed jars or cans on rack in the canner. Allow space for steam to flow around each container. If there are two layers of cans or jars, stagger the top layer. Use a rack between layers of jars.
- Fasten canner cover securely so that all steam escapes through the petcock or weighted-gage opening.
- Let steam pour steadily from vent for 10 minutes to drive all air from the canner. Then close petcock or put on weighted gage.
- Let pressure rise to 10 pounds (240° F.). The moment this pressure is reached, start to count processing time. Regulate heat under the canner to maintain even pressure. Do not lower pressure by opening petcock. Keep drafts from blowing on canner. Fluctuating pressure during processing causes liquid to be drawn out of glass jars.
- Watch processing time carefully. When time is up, remove canner from heat immediately.
- If meat is packed in jars, let canner stand until pressure drops to zero. Do not pour cold water over canner. When pressure is reduced suddenly, jars lose liquid. After pressure registers zero, wait a minute or two. Then slowly open petcock or take off weighted gage. Unfasten cover and tilt the far side up so steam escapes away from you. Take jars from the canner.
- If meat is packed in cans, remove canner from heat as soon as processing time is up. Open petcock or take off weighted gage at once to release steam. Then unfasten cover, tilting far side up so steam escapes away from your face. Remove cans.

Yield of Canned Meat From Fresh

The number of jars or cans you get from a given amount of raw meat varies with the size of the pieces and the way the meat is packed.

For a 1-quart jar, allow approximately the following amounts of fresh, untrimmed meat with bone or ready-to-cook chicken:

	Pounds
Beef:	
Round	3 to 3½
Rump	5 to 5½
Pork loin	5 to 5½
Chicken:	
Canned with bone	3½ to 4¼
Canned without bone	5½ to 6¼

After-Canning Jobs

Cooling

Glass jars

As soon as you take jars out of the canner, complete seals if necessary.

Cool jars top side up. Put them on a rack or folded cloth to cool. Keep them away from drafts. Don't cover.

When jars are cool, take off screw bands. Do not force bands that stick; loosen by covering them with a hot damp cloth. Wash bands and store them in a dry place.

Tin cans

As soon as you take cans out of the canner, put them in cold water. Change water frequently for fast cooling. Remove cans from water while they are still warm so they will dry in the air. If you stack cans, stagger them to allow air circulation.

Checking Seals

Check containers for leaks when jars or cans are thoroughly cool.

Occasionally, a can packed too full bulges at the ends. Set this aside and use within a few days. This will prevent later confusion with cans that bulge from spoilage during storage.

On the day after canning, examine each jar carefully. Turn it partially over. If jar has a flat metal lid, test seal by tapping center of lid with a spoon. A clear ring means a good seal. A dull note, however, does not always mean a poor seal. Another test is to press on the center of the lid; jar is sealed if lid is down and does not move.

Examine seams and seals carefully on all cans. Can ends should be almost flat, with a slight inward curve. Buckled or broken seams may be caused by cooling cans too fast or by not filling cans with enough meat.

Do not store leaky jars and cans. Either use the food at once or can it again in another container. Begin the second canning by heating meat through. Then pack and process it in a pressure canner for the full time recommended.

Do not open and refill jars that have lost liquid during processing. Loss of liquid does not cause canned meat to spoil. Opening would contaminate the sterile contents, and meat would have to be processed again to insure safety.

Labeling

Wipe containers after they are cool. Label each jar and can to show contents and date of canning. If you canned more than one lot on one day, add a lot number.

Storing

Select a cool, dry place for storing canned meat and poultry. Protect meat from heat, from freezing, and from dampness.

Heat causes canned foods to lose quality. Do not store canned meats in direct sunlight, near hot pipes, or near heat.

Freezing does not cause canned meat to spoil, but it may damage the seal so that spoilage begins. In an unheated storage area, cover jars and cans with a clean, old blanket or wrap them in newspapers.

Dampness may corrode cans or metal jar lids and cause leakage.

Spoilage

Immediately destroy any canned meat that has spoiled. Burn it or dispose of it where it cannot be eaten by humans or animals.

Do not taste canned meat that you suspect of being spoiled.

Take these positive steps to insure the safety of canned meat—

- Keep meat and equipment clean.
- Follow recommended methods, including processing times and temperatures.
- Cool and store properly.

To avoid any risk of botulism—a serious food poisoning—it is essential that the pressure canner be in perfect order and that every canning recommendation be followed exactly. Unless you are absolutely sure of your gage and canning methods, boil home-canned meat 20 minutes in a covered pan before tasting or using.

Boiling is the best way to find out if canned meat is safe. Heat brings out the characteristic odor of spoiled meat. If meat develops such an odor, destroy it without tasting.

If boiled meat is not to be used at once, or if it is to be used in salads or sandwiches, refrigerate it immediately.

Be alert to signs of spoilage when you take meat containers from storage. Bulging jar lids or rings, gas bubbles, leaks, bulging can ends—these may mean the seal has broken and the food has spoiled. Test each can by pressing the ends; ends should not bulge or snap back.

Check the contents as you open the container. Spurting liquid, off-odor, and color changes in meats are danger signals.

Sulfur in meat often causes metal lids or cans to darken. This discoloration does not affect the safety of the meat.

How To Can MEAT—raw pack

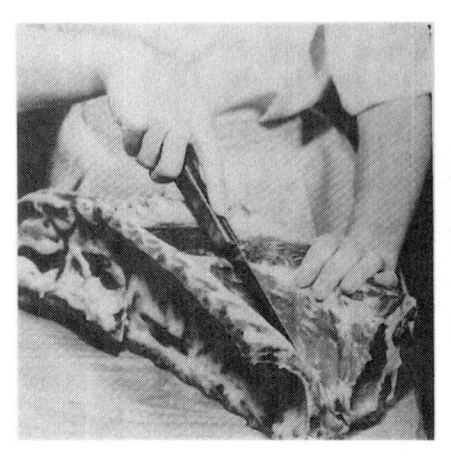

1. Cut meat carefully from bone. Trim away most of fat without unduly slashing the lean part of meat.

2. Cut meat in jar-length pieces, so grain of meat runs length of jar. Fill jars to 1 inch of top with one or more pieces of meat.

3. Set open, filled jars on rack in pan of boiling water. Keep water level 2 inches below jar tops. Insert thermometer in center of a jar (above), cover pan, and heat meat slowly to 170° F. Without thermometer, cover pan; heat slowly for 75 minutes.

4. Remove jars from pan. Add salt if desired. Wipe jar rim clean. Place lid so that sealing compound is next to glass (above). Screw the metal band down tight by hand. When band is screwed tight, this lid has enough "give" to let air escape during processing.

5. Have 2 or 3 inches of boiling water in pressure canner—enough to keep it from boiling dry during processing. Put jars in canner (above); fasten cover. Let steam pour from open petcock or weighted-gage opening 10 minutes. Shut petcock or put on gage.

6. When pressure reaches 10 pounds, note time. Adjust heat under canner to keep pressure steady. Process pint jars packed with large pieces of meat 75 minutes; process quart jars 90 minutes. When processing time is up, slide canner away from heat.

7. Let pressure fall to zero (30 minutes). Wait a minute or two, then slowly open petcock. Unfasten cover, tilting far side up to keep steam away from your face.

8. Set jars on rack to cool overnight. Keep them away from drafts, but do not cover. When jars are thoroughly cool, remove metal bands and wipe jars clean. Label and store.

Directions for canning cut-up meat by hot-pack and raw-pack methods begin on page 18.

Home Canning of Meat and Poultry (15) 47

How To Can CHICKEN—hot pack

1. Rinse and drain (p. 4); then use a sharp knife to disjoint bird. Pull on leg or wing as you cut through the joint.

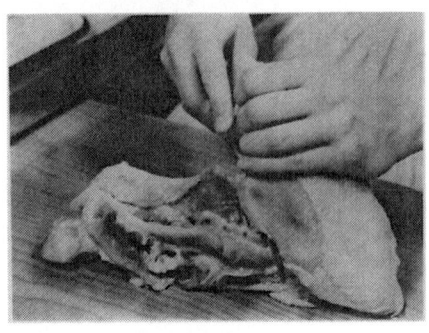

2. Cut from end of breastbone to backbone along ends of ribs. Separate breast and back. Break backbone; cut back in half.

3. Cut breast straight down between wishbone and point of breast. Leave meat on wishbone.

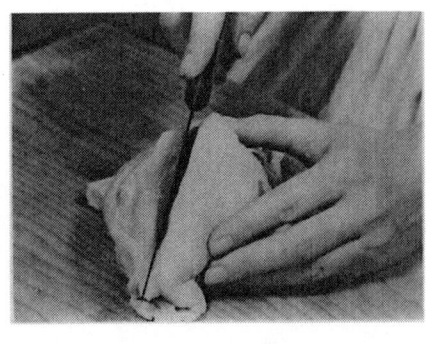

4. Remove breast meat from center bone by carving down the bone on one side of breast. Repeat on other side of breastbone.

5. Cut legs into drumsticks and thighs. Saw drumsticks off short, if desired. Sort into meaty and bony pieces; set aside giblets to can separately.

6. Pour enough hot water or broth (p. 8) over raw meaty pieces in pan to cover meat. Put on lid; precook meat to medium done (when cut at center, pieces show almost no pink color).

7. Pack hot chicken loosely into jars. Place thighs and drumsticks with skin next to glass; breasts in center of jar; smaller pieces fitted in. Leave 1 inch at the top of jar. Add salt, if desired. Cover chicken with boiling broth. Again leave 1 inch of space at top of jar.

8. Wipe jar rim clean. Place lid with sealing compound next to glass. Screw metal band tight. Have 2 or 3 inches of boiling water in pressure canner to prevent it from boiling dry during processing. Place jars in canner (left) and fasten lid securely. Let steam pour from open petcock or weighted-gage opening for 10 minutes. Then shut petcock or put on weighted gage.

9. When pressure is 10 pounds, note time. Adjust heat to keep pressure steady. Process pint jars of chicken with bone 65 minutes; quarts, 75 minutes. Slide canner off heat when time is up. Let pressure fall to zero (about 30 minutes). Wait a minute or two. Open petcock slowly. Unfasten cover, tilting far side up to keep steam away from your face. Cool jars overnight. Wipe clean; label (left). Before storing canned chicken, remove screw bands.

Directions for canning poultry by hot-pack and raw-pack methods begin on page 21.

Home Canning of Meat and Poultry (17) 49

Directions for Meat

Directions for canning cut-up meat may be used for beef, veal, pork, lamb, and mutton. Meat from large-game animals may be canned by the same directions.

Use tender meat—loin and cuts suitable for roasts, steaks, and chops—for canning as large pieces. Use less tender cuts that contain more connective tissue and small pieces for canning as stew meat or ground meat. Use bony pieces for soup.

Cut-Up Meat

Follow directions for cutting up meat (p. 14).

Cut tender meat into jar- or can-length strips. Strips should slide into jars or cans easily, with the grain of the meat running the length of the container. Strips may be any convenient thickness, from 1 or 2 inches to jar or can width.

Cut less tender meat into chunks or small pieces suitable for stew meat.

Small, tender pieces may be packed by themselves, with meat strips, or with stew meat.

Hot pack

Put meat in large shallow pan; add just enough water to keep from sticking. Cover pan. Precook meat slowly until medium done. Stir occasionally, so meat heats evenly.

Glass jars.—Pack hot meat loosely. Leave 1 inch of space at top of jars. Add salt if desired: ½ teaspoon to pints or 1 teaspoon to quarts. Cover meat with boiling meat juice, adding boiling water if needed. Leave 1 inch of space at top of jars. Adjust lids. Process in a pressure canner at 10 pounds pressure (240° F.)—

Pint jars	75 minutes
Quart jars	90 minutes

Tin cans.—Pack hot meat loosely. Leave ½ inch of space above meat. Add salt if desired: ½ teaspoon to No. 2 cans or ¾ teaspoon to No. 2½ cans. Fill cans to top with boiling meat juice, adding boiling water if needed. Seal. Process in a pressure canner at 10 pounds pressure (240° F.)—

No. 2 cans	65 minutes
No. 2½ cans	90 minutes

Raw pack

Cut up meat (p. 14). Pack containers loosely with raw, lean meat.

Glass jars.—Leave 1 inch of space above meat. To exhaust air, cook raw meat in jars at slow boil to 170° F., or until medium done (about 75 minutes). (See p. 9.) Add salt if desired: ½ teaspoon per pint or 1 teaspoon per quart. Adjust lids. Process in a pressure canner at 10 pounds pressure (240° F.)—

Pint jars	75 minutes
Quart jars	90 minutes

Tin cans.—Pack tin cans to top. To exhaust air, cook raw meat in cans at slow boil to 170° F., or until medium done (about 50 minutes). (See p. 9.) Press meat down ½ inch below rim, and add boiling water to fill to top, if needed. Add salt if desired: ½ teaspoon to No. 2 cans or ¾ teaspoon to No. 2½ cans. Seal cans. Process in a pressure canner at 10 pounds pressure (240° F.)—

    No. 2 cans. . . . . . . 65 minutes
    No. 2½ cans. . .   90 minutes

Ground Meat

For grinding, start with fresh, clean, cold meat. Use small pieces of meat from less tender cuts.

Never mix leftover scraps with fresh meat. Don't use lumps of fat.

If desired, add 1 level teaspoon of salt per pound of ground meat. Mix well.

Hot pack

Shape ground meat into fairly thin patties that can be packed into jars or cans without breaking.

Precook patties in slow oven (325° F.) until medium done. (When cut at center, patties show almost no red color.) Skim fat off drippings; do not use fat in canning.

Glass jars.—Pack patties, leaving 1 inch of space above meat. Cover with boiling meat juice to 1 inch of top of jars. Adjust jar lids. Process in a pressure canner at 10 pounds pressure (240° F.)—

    Pint jars. . . . . 75 minutes
    Quart jars. . . .   . . . 90 minutes

Tin cans.—Pack patties to ½ inch of top of cans. Cover with boiling meat juice to fill cans to top; seal.

Process in a pressure canner at 10 pounds pressure (240° F.)—

    No. 2 cans. .. . ... 65 minutes
    No. 2½ cans.. . .... 90 minutes

Raw pack

Raw pack is suitable for tin cans. Ground meat canned in bulk is difficult to get out of jars.

Tin cans.—Pack raw ground meat solidly to the top of the can. To exhaust air, cook meat at slow boil to 170° F., or until medium done (about 75 minutes). (See p. 9.) Press meat down into cans ½ inch below rim. Seal. Process in a pressure canner at 10 pounds pressure (240° F.)—

    No. 2 cans. . . . . . . 100 minutes
    No. 2½ cans. . . . . . 135 minutes

Sausage

Hot pack

Use any tested sausage recipe.

Use seasonings sparingly because sausage changes flavor in canning and storage. Measure spices, onion, and garlic carefully. Omit sage—it makes canned sausage bitter.

Shape sausage meat into patties. Precook, pack, and process as directed for hot-packed ground meat.

Corned Beef

Hot pack

Use any tested recipe to make corned beef.

Wash corned beef. Drain. Cut in pieces or strips that fit in containers.

Cover meat with cold water and bring to a boil. If broth is very salty, drain meat; boil again in fresh water. Pack while hot.

Glass jars.—Leave 1 inch of space above meat. Cover meat with boiling broth or boiling water. Leave 1 inch of space at top of jars. Adjust lids. Process in a pressure canner at 10 pounds pressure (240° F.)—

    Pint jars............ 75 minutes
    Quart jars.......... 90 minutes

Tin cans.—Leave ½ inch of space above meat. Fill cans to top with boiling broth or boiling water. Seal. Process in a pressure canner at 10 pounds pressure (240° F.)—

    No. 2 cans......... 65 minutes
    No. 2½ cans....... 90 minutes

Meat-Vegetable Stew

Raw pack

Beef, lamb, or veal, cut in 1½-inch
  cubes........................... 2 quarts
Potatoes, pared or scraped, cut in
  ½-inch cubes.................. 2 quarts
Carrots, pared or scraped, cut in
  ½-inch cubes.................. 2 quarts
Celery, ¼-inch pieces............. 3 cups
Onions, small whole, peeled..... 7 cups

Combine ingredients. Yield is 7 quarts or 16 pints.

Glass jars.—Fill jars to top with raw meat-vegetable mixture. Add salt if desired: ½ teaspoon per pint or 1 teaspoon per quart. Adjust lids. Process in a pressure canner at 10 pounds pressure (240° F.)—

    Pint jars........ 60 minutes
    Quart jars...... 75 minutes

Tin cans.—Fill cans to top with raw meat-vegetable mixture. Do not add liquid. Add salt if desired: ½ teaspoon to No. 2 cans or 1 teaspoon to No. 2½ cans. To exhaust air, cook stew at slow boil to 170° F., or until medium done (about 50 minutes). (See p. 9.) Seal cans. Process in a pressure canner at 10 pounds pressure (240° F.)—

    No. 2 cans........ 40 minutes
    No. 2½ cans...... 45 minutes

Heart and Tongue

Hot pack

Heart and tongue usually are served as fresh meat. To can, prepare as described below; then follow hot pack directions (p. 18).

Heart.—Remove thick connective tissue before cutting into pieces.

Tongue.—Drop tongue into boiling water and simmer about 45 minutes, or until skin can be removed. Then cut into pieces.

Soup Stock

Hot pack

For canning, make meat stock fairly concentrated. Cover bony pieces of meat (or chicken) with lightly salted water. Simmer until tender.

Skim off fat. Remove all bones. Leave meat and sediment in stock.

Glass jars.—Pour boiling soup stock into jars, leaving 1 inch of space at top. Adjust lids. Process in a pressure canner at 10 pounds pressure (240° F.)—

    Pint jars........ 20 minutes
    Quart jars...... 25 minutes

Tin cans.—Fill cans to top with boiling soup stock. Seal. Process in a pressure canner at 10 pounds pressure (240° F.)—

    No. 2 cans........ 20 minutes
    No. 2½ cans...... 25 minutes

Directions for Poultry

Directions for poultry may be used to can chicken, duck, goose, guinea, squab, and turkey. These directions also apply to game birds.

Domestic rabbits and small-game animals should be canned like poultry.

Poultry, rabbits, and small-game animals may be canned with or without bone.

To make soup stock from poultry for canning, follow directions for meat.

Cut-Up Poultry

Follow directions for cutting up poultry (p. 16). Sort into meaty and bony pieces. Use bony pieces for broth (p. 8) or soup (p. 20). Set aside giblets to can separately.

Hot pack, with bone

Bone breast. Saw drumsticks off short. Leave bone in other meaty pieces. Trim off large lumps of fat.

Place raw meaty pieces in pan and cover with hot broth or water. Put on lid. Heat, stirring occasionally until medium done. To test, cut piece at center; if pink color is almost gone, meat is medium done.

Pack poultry loosely. Place thighs and drumsticks with skin next to glass or tin. Fit breasts into center and small pieces where needed.

Glass jars.—Pack jars, leaving 1 inch of space above poultry. Add salt if desired: ½ teaspoon per pint or 1 teaspoon per quart. Cover poultry with boiling broth, leaving 1 inch of space at top of jar. Adjust jar lids. Process in pressure canner at 10 pounds pressure (240° F.)—

    Pint jars.......... 65 minutes
    Quart jars......... 75 minutes

Tin cans.—Pack cans, leaving ½ inch of space above poultry. Add salt if desired: ½ teaspoon to No. 2 cans or ¾ teaspoon to No. 2½ cans. Fill cans to top with boiling broth. Seal. Process in a pressure canner at 10 pounds pressure (240° F.)—

    No. 2 cans........ 55 minutes
    No. 2½ cans....... 75 minutes

Hot pack, without bone

Cut up poultry (p. 16). Remove bone—but not skin—from meaty pieces either before or after precooking.

Glass jars.—Pack jars loosely with hot poultry, leaving 1 inch of space above poultry at top of jars. Add salt if desired: ½ teaspoon per pint or 1 teaspoon per quart. Pour in boiling broth; leave 1 inch of space at top of jar. Adjust jar lids. Process in a pressure canner at 10 pounds pressure (240° F.)—

    Pint jars.......... 75 minutes
    Quart jars...... 90 minutes

Tin cans.—Pack loosely, leaving ½ inch above poultry. Add salt if desired: ½ teaspoon to No. 2 cans or ¾ teaspoon to No. 2½ cans. Fill cans to top with boiling broth. Seal.

Process in a pressure canner at 10 pounds pressure (240° F.)—

    No. 2 cans....... 65 minutes
    No. 2½ cans....... 90 minutes

Raw pack, with bone

Cut up poultry (see p. 16).

Bone breast. Saw drumsticks off short. Leave bone in other meaty pieces. Trim off large lumps of fat.

Pack raw poultry loosely. Place thighs and drumsticks with skin next to glass or tin. Fit breasts into center and small pieces where needed.

Glass jars (air exhausted).—Pack jars to 1 inch of top. To exhaust air, cook raw poultry in jars at slow boil to 170° F., or until medium done (about 75 minutes). (See p. 9.) Add salt if desired: ½ teaspoon per pint or 1 teaspoon per quart. Adjust lids. Process in a pressure canner at 10 pounds pressure (240° F.)—

    Pint jars........ 65 minutes
    Quart jars........ 75 minutes

Glass jars (air not exhausted).—Fill jars loosely with raw pieces of poultry to 1 inch of top. Do not exhaust. Add salt if desired: ½ teaspoon per pint or 1 teaspoon per quart. Adjust lids. Process in a pressure canner at 10 pounds pressure (240° F.)—

    Quart jars....... 80 minutes

Tin cans.—Pack cans to top. To exhaust air, cook raw poultry in cans at slow boil to 170° F., or until medium done (about 50 minutes). (See p. 9.) Add salt if desired: ½ teaspoon to No. 2 cans or ¾ teaspoon to No. 2½ cans. Seal cans. Process in a pressure canner at 10 pounds pressure (240° F.)—

    No. 2 cans....... 55 minutes
    No. 2½ cans....... 75 minutes

Raw pack, without bone

Cut up poultry (p. 16). Remove bone—but not skin—from meaty pieces before packing containers.

Glass jars.—Pack raw poultry in jars to 1 inch of top. To exhaust air, cook poultry in jars at slow boil to 170° F., or until medium done (about 75 minutes). (See p. 9.) Add salt if desired: ½ teaspoon per pint or 1 teaspoon per quart. Adjust lids. Process in a pressure canner at 10 pounds pressure (240° F.)—

    Pint jars.......... 75 minutes
    Quart jars........ 90 minutes

Tin cans.—Pack raw poultry to top of cans. To exhaust air, cook poultry in cans at slow boil to 170° F., or until medium done (about 50 minutes). (See p. 9.) Add salt if desired: ½ teaspoon to No. 2 cans or ¾ teaspoon to No. 2½ cans. Seal cans. Process in a pressure canner at 10 pounds pressure (240° F.)—

    No. 2 cans....... 65 minutes
    No. 2½ cans....... 90 minutes

Giblets

Use pint jars or No. 2 cans.

Wash and drain giblets.

Pack gizzards and hearts together. Precook and pack livers separately to avoid blending of flavors.

Hot pack

Put giblets in pan; cover with hot broth or hot water. Cover pan and precook giblets until medium done. Stir occasionally. Pack hot.

Glass jars.—Leave 1 inch of space above giblets. Add boiling broth or boiling water, leaving 1 inch of space below jar tops. Adjust lids. Process in a pressure canner at 10 pounds pressure (240° F.)—

Pint jars............ 75 minutes

Tin cans.—Leave one-half inch of space above giblets. Fill cans to top with boiling broth or boiling water. Seal. Process in a pressure canner at 10 pounds pressure (240° F.)—

No. 2 cans........ 65 minutes

Questions and Answers

Q. Why must a pressure canner be used for canning meat and poultry?
A. To insure a safe product. It takes a combination of high temperature and sufficient processing time to make sure of killing bacteria that cause dangerous spoilage in canned meat and poultry. The only practical way to get the necessary high temperature is to use a pressure canner.

Q. How should meat and poultry for canning be handled?
A. Keep meat and poultry clean and sanitary. Chill at once and keep cold until canning time. (See p. 4.)

Q. Why is liquid sometimes lost from glass jars during processing?
A. Loss of liquid may be due to packing jars too full, fluctuating pressure in a pressure canner, or lowering pressure too suddenly.

Q. Should liquid lost during processing be replaced?
A. No. Loss of liquid does not cause meat to spoil, although the meat above the liquid may darken. Never open a jar and refill with liquid—this would let in bacteria and meat would have to be processed again.

Q. Is it safe to can meat and poultry without salt?
A. Yes. Salt is used for flavor only and is not necessary for safe processing.

Q. Is it safe to leave food in tin cans after opening?
A. Yes. Food in tin cans needs only to be covered and refrigerated.

Q. Is it all right to use preservatives in home canning?
A. No. Some canning powders or other chemical preservatives may be harmful.

Q. Should processing times be changed for different types of ranges?
A. No. Processing times and temperatures given in this bulletin are for canning in a pressure canner and may be used for any type of range.

Q. Is it possible to can frozen meat or poultry?
A. Yes, frozen meat or poultry may be canned. (For directions, see p. 4.)

Index

	Page
Altitude, high, canning at	5
Beef	18
Beef-vegetable stew	20
Botulism	4, 13
Broth	8
Canning methods	8
Chilling meat	4
Corned beef	19
Discoloration	13
Equipment	4
Fat	9
Frozen meat, canning	4
Gage	5
Giblets	22
Glass jars:	
Cooling	12
Closures	6, 10
Exhausting	9
Packing	9
Processing	11
Use	6
Heart	20
Labeling	12
Lamb	18
Lamb-vegetable stew	20
Large game	18
Liquid:	
Loss	9, 23
Use	9
Meat, cut-up:	
Hot pack	18
Raw pack	18
Meat, ground:	
Hot pack	19
Raw pack	19
Meat-vegetable stew	20
Mutton	18

	Page
Packing:	
Hot pack	9
Raw pack	9
Pork	18
Poultry, cut-up:	
Hot pack, with bone	21
Hot pack, without bone	21
Raw pack, with bone	22
Raw pack, without bone	22
Pressure canner:	
Care	5
Use	5, 11
Pressure saucepan	5
Rabbit	21
Rubber rings	7
Salt	9, 23
Sausage	19
Sealer, tin-can	7, 10
Seals, testing	7
Small game	21
Soup stock	20
Spoilage	4, 5, 13
Stew	20
Storing	13
Thermometer	7
Tin cans:	
Cooling	12
Exhausting	9
Packing	9
Processing	11
Sealing	10
Types	7
Use	7
Tongue	20
Veal	18
Veal-vegetable stew	20
Yields of canned meat from fresh	11

How To Make Jellies, Jams, and Preserves at Home

Jelly, jam, conserve, marmalade, preserves—any of these fruit products can add zest to meals. Most of them also provide a good way to use fruit not at its best for canning or freezing—the largest or smallest fruits and berries and those that are irregularly shaped.

Basically these products are much alike; all of them are fruit preserved by means of sugar, and usually all are jellied to some extent. Their individual characteristics depend on the kind of fruit used and the way it is prepared, the proportions of different ingredients in the mixture, and the method of cooking.

Jelly is made from fruit juice; the product is clear and firm enough to hold its shape when turned out of the container. Jam, made from crushed or ground fruit, tends to hold its shape but generally is less firm than jelly. Conserves are jams made from a mixture of fruits, usually including citrus fruit; often raisins and nuts are added. Marmalade is a tender jelly with small pieces of fruit distributed evenly throughout; a marmalade commonly contains citrus fruit. Preserves are whole fruits or large pieces of fruit in a thick sirup, often slightly jellied.

Not all fruits have the properties needed for making satisfactory jellied products, but with the commercial pectins now on the market, the homemaker need not depend on the jellying quality of the fruit for successful results.

This publication tells how to make various kinds of jellies, jams, and conserves, with and without added pectin. It also includes recipes for marmalades and preserves made with no added pectin.

Four essential ingredients

Proper amounts of fruit, pectin, acid, and sugar are needed to make a jellied fruit product.

Fruit

Fruit gives each product its characteristic flavor and furnishes at least part of the pectin and acid required for successful gels.

Flavorful varieties of fruits are best for jellied products because the fruit flavor is diluted by the large proportion of sugar necessary for proper consistency and good keeping quality.

Pectin

Some kinds of fruit have enough natural pectin to make high-quality products. Others require added pectin, particularly when they are used for making jellies, which should be firm enough to hold their shape. All fruits have more pectin when they are underripe.

Commercial fruit pectins, which are made from apples or citrus fruits, are on the market in two forms—liquid and powdered. Either form is satisfactory when used in a recipe developed especially for that form.

These pectins may be used with any fruit. Many homemakers prefer the added-pectin method for making jellied fruit products because fully ripe fruit can be used, cooking time is shorter and is standardized so that there is no question when the product is done, and the yield from a given amount of fruit is greater.

Fruit pectins should be stored in a cool, dry place so they will keep their gel strength. They should not be held over from one year to the next.

Acid

Acid is needed for flavor and for gel formation. The acid content varies in different fruits and is higher in underripe fruits.

With fruits that are low in acid, lemon juice or citric acid is commonly added in making jellied products. Also, commercial fruit pectins contain some acid.

In the recipes in this publication, lemon juice is included to supply additional acid when necessary. If you wish, use ⅛ teaspoon of crystalline citric acid for each tablespoon of lemon juice called for.

Sugar

Sugar helps in gel formation, serves as a preserving agent, and contributes to the flavor of the jellied product. It also has a firming effect on fruit, a property that is useful in the making of preserves.

Beet and cane sugar can be used with equal success. Although they come from different sources, they have the same composition.

Equipment and containers

Equipment

A large kettle is essential. To bring mixture to a full boil without boiling over, use an 8- or 10-quart kettle with a broad flat bottom.

A jelly bag or a fruit press may be used for extracting fruit juice for jellies. The bag may be made of several thicknesses of closely woven cheesecloth, of firm unbleached muslin, or canton flannel with napped side in. Use a jelly bag or cheesecloth to strain pressed juice. A special stand or collander will hold the jelly bag.

A jelly, candy, or deep-fat thermometer is an aid in making fruit products without added pectin.

Other kitchen equipment that may be useful includes a quart measure, measuring cup and spoons, paring and utility knives, food chopper, masher, reamer, grater, bowls, wire basket, colander, long-handled spoon, ladle, clock with second hand, and household scale.

Containers

Jelly glasses or canning jars may be used as containers for jellied fruit products. Be sure all jars and closures are perfect. Discard any with cracks or chips; defects prevent airtight seals.

For jellies to be sealed with paraffin, use glasses or straight-sided containers that will make an attractive mold.

For jams, preserves, conserves, and marmalades, use canning jars with lids that can be tightly sealed and processed. Paraffin tends to loosen and break the seal on these products.

Get glasses or jars ready before you start to make the jellied product. Wash containers in warm, soapy water and rinse with hot water. Sterilize jelly containers in boiling water for 10 minutes. Keep all containers hot—either in a slow oven or in hot water—until they are used. This will prevent containers from breaking when filled with hot jelly or jam.

Wash and rinse all lids and bands. Metal lids with sealing compound may need boiling or holding in boiling water for a few minutes—follow the manufacturer's directions. Use new lids; bands and jars may be reused.

If you use porcelain-lined zinc caps, have clean, new rings of the right size for jars. Wash rings in hot, soapy water. Rinse well.

Making and storing jellied fruit products

Directions for making different kinds of jellied fruit products are given in this publication. The formulas selected take into account the natural pectin and acid content of the specified fruit.

Tips on jellied fruit products

For freshness of flavor. To have jellied fruit products at their best, make up only the quantity that can be used within a few months; they lose flavor in storage.

For softer or firmer products. If fruit with average jellying properties is used, the jellied products made according to directions in this publication should be medium firm for their type. However, because various lots of fruit differ in composition, it is not possible to develop formulas that will always give exactly the same results.

If the first batch from a particular lot of fruit is too soft or too firm, adjust the proportions of fruit or the cooking time for the next batch.

In products made with added pectin—

For a softer product, use ¼ to ½ cup *more* fruit or juice.

For a firmer product, use ¼ to ½ cup *less* fruit or juice.

In products made without added pectin—

For a softer product, *shorten* the cooking time.

For firmer product, *lengthen* the cooking time.

Using canned, frozen, or dried fruits. Any fresh fruit may be canned or frozen as fruit or juice and used in jellied products later. Both fruit and juice should be canned or frozen unsweetened; if sweetened—the amount of sugar should be noted and subtracted from the amount in the jelly or jam recipe. Fruit should be canned in its own juice or with only a small amount of water. If you plan to use canned or frozen fruit without added pectin, it is best to use part underripe fruit, especially for jelly.

Unsweetened commercially canned or frozen fruit or juice can also be used in jellied products. Concentrated frozen juices make very flavorful jellies. Commercially canned or frozen products are made from fully ripe fruit, and require added pectin if used for jelly.

Dried fruit may be cooked in water until tender and used to make jams and conserves, with or without added pectin as required.

Filling and sealing containers

Prepare canning jars and lids or jelly glasses as directed (p. 3).

To seal with lids.—Use only standard home canning jars. For jars with two-piece lids: Use new lids; bands may be reused. Fill hot jars to ⅛ inch of top with hot jelly or fruit mixture. Wipe jar rim clean, place hot metal lid on jar with sealing compound next to glass, screw metal band down firmly, and stand jar upright to cool. For jars with porcelain-lined zinc caps: Place wet rubber ring on shoulder of empty jar. Fill jar to ⅛ inch of top, screw cap down tight to complete seal, and stand jar upright to cool.

Work quickly when packing and sealing jars. To keep fruit from floating to the top, gently shake jars of jam occasionally as they cool.

To seal with paraffin.—Use this method only with jelly. Pour hot jelly mixture immediately into hot containers to within ½ inch of top. Cover with hot paraffin. Use only enough paraffin to make a layer ⅛ inch thick. A single thin layer—which can expand or contract readily—gives a better seal than one thick layer or two thin layers. Prick air bubbles in paraffin. Bubbles cause holes as paraffin hardens; they may prevent a good seal. A double boiler is best for melting paraffin and keeping it hot without reaching smoking temperature.

To process jams, conserves, marmalades, and preserves.—Processing of these products is recommended in warm or humid climates. Inexpensive enamelware canners may be purchased at most hardware or variety stores. However, any large metal container may be used if it—

- Is deep enough to allow for 1 or 2 inches of water above the tops of the jars, plus a little extra space for boiling.
- Has a close-fitting cover.
- Has a wire or wood rack with partitions to keep jars from touching each other or the bottom or sides of the container.

Put filled home canning jars into a water bath canner or a container filled with hot water. Add hot water if needed to bring water an inch or two over tops of jars. Bring water to a rolling boil and boil gently for 5 minutes.

Remove jars from canner after processing. Cool away from drafts before storing.

Storing jellied fruit products

Let products stand undisturbed overnight to avoid breaking gel. Cover jelly glasses with metal or paper lids. Label with name, date, and lot number if you make more than one lot a day. Store in a cool, dry place; the shorter the storage time, the better the eating quality of the product.

Uncooked jams (p. 26) may be held up to 3 weeks in a refrigerator; for longer storage, they should be placed in a freezer.

Jellies

Jelly is clear and bright with the natural color and flavor of the fruit from which it is made. It is tender yet firm enough to hold its shape when cut.

When making jelly, with or without added pectin, it is best to prepare small cooking lots, as indicated in the recipes that follow. Increasing the quantities given is not recommended.

To prepare fruit

Approximate amounts of fruits needed to yield the amount of juice called for are given in each recipe. However, the exact amount will vary with juiciness of the particular lot of fruit used.

Wash all fruits in cold running water, or wash them in several changes of cold water, lifting them out of the water each time. Do not let fruit stand in water.

Prepare fruit for juice extraction as directed in the recipe; the method differs with different kinds of fruit. Juicy berries may be crushed and the juice pressed out without heating. For firm fruits, heating is needed to help start the flow of juice, and usually some water is added when the fruit is heated.

To extract juice

Put the prepared fruit in a damp jelly bag, fruit press, or a double layer of damp cheesecloth to extract the juice. The clearest jelly comes from juice that has dripped through a jelly bag without pressing. But a greater yield of juice can be obtained by twisting the bag of fruit tightly and squeezing or pressing, or by using a fruit press. Pressed juice should be re-strained through a double thickness of damp cheesecloth or a damp jelly bag; the cloth or bag should not be squeezed.

To make jelly

With added pectin. In this publication some of the recipes have been developed with powdered pectin, others with liquid pectin. Because of differences between the two forms, each should be used only in recipes worked out for that form.

The order in which the ingredients are combined depends on the form of pectin. Powdered pectin is mixed with the unheated fruit juice. Liquid pectin is added to the boiling juice and sugar mixture.

Boiling time is the same with either form of pectin; a 1-minute boiling period is recommended. Accurate timing is important. Time should not be counted until the mixture has reached a full rolling boil—one that cannot be stirred down.

For best flavor, use fully ripe fruit when making jelly with added pectin.

Without added pectin. Jellies made without added pectin require less sugar per cup of fruit juice than do those with added pectin, and longer boiling is necessary to bring the mixture to the proper sugar concentration. Thus the yield of jelly per cup of juice is less.

It is usually best to have part of the fruit underripe when no pectin is added, because underripe fruit has a higher pectin content. The use of one-fourth underripe and three-fourths fully ripe fruit is generally recommended to assure sufficient pectin for jelly.

To test for pectin in fruit juice

A rough estimate of the amount of pectin in fruit juice may be obtained through use of denatured alcohol or a jelmeter.

Alcohol test. Add 1 tablespoon cooked, cooled fruit juice to 1 tablespoon denatured alcohol. Stir slightly to mix. Juices rich in pectin will form a solid jelly-like mass. Juices low in pectin will form small particles of jelly-like material.

NOTE: *Denatured alcohol is poisonous. Do not taste the tested juice. Wash all utensils used in this test thoroughly.*

Jelmeter test. A jelmeter is a graduated glass tube with an opening at each end. The rate of flow of fruit juice through this tube gives a rough estimate of the amount of pectin in the juice.

If a test indicates that the juice is low in pectin, use a recipe calling for the addition of powdered or liquid pectin.

To test for doneness

The biggest problem in making jelly without added pectin is to know when it is done. It is particularly important to remove the mixture from the heat before it is overcooked. Although an undercooked jelly can sometimes be recooked to make a satisfactory product (see p. 32), there is little that can be done to improve an overcooked mixture. Signs of overcooking are a change in color of mixture and a taste or odor of caramelized sugar.

Three methods that may be used for testing doneness of jelly made at home are described below. Of these, the temperature test probably is the most dependable.

Temperature test. Before cooking the jelly, take the temperature of boiling water with a jelly, candy, or deep-fat thermometer. Cook the jelly mixture to a temperature 8° F. higher than the boiling point of water. At that point the concentration of sugar will be such that the mixture should form a satisfactory gel.

It is necessary to find out at what temperature water boils in your locality because the boiling point differs at different altitudes. And, because the boiling point at a given altitude may change with different atmospheric conditions, the temperature of boiling water should be checked shortly before the jelly is to be made.

For an accurate thermometer reading, have the thermometer in a vertical position and read it at eye level. The bulb of the thermometer must be completely covered with the jelly mixture, but must not touch the bottom of the kettle.

Spoon or sheet test. Dip a cool metal spoon in the boiling jelly mixture. Then raise it at least a foot above the kettle, out of the steam, and turn the spoon so the sirup runs off the side. If the sirup forms two drops that flow together and fall off the spoon as one sheet, the jelly should be done. This test has been widely used; however, it is not entirely dependable.

Refrigerator test. Pour a small amount of boiling jelly on a cold plate, and put it in the freezing compartment of a refrigerator for a few minutes. If the mixture gels, it should be done. During this test, the jelly mixture should be removed from the heat.

Apple Jelly

without added pectin

4 cups apple juice (about 3 pounds apples and 3 cups water)
2 tablespoons strained lemon juice, if desired
3 cups sugar

- **To prepare juice.** Select about one-fourth underripe and three-fourths fully ripe tart apples. Sort, wash, and remove stem and blossom ends; do not pare or core. Cut apples into small pieces. Add water, cover, and bring to boil on high heat. Reduce heat and simmer for 20 to 25 minutes, or until apples are soft. Extract juice (p. 6).
- **To make jelly.** Measure apple juice into a kettle. Add lemon juice and sugar and stir well. Boil over high heat to 8° F. above the boiling point of water, or until jelly mixture sheets from a spoon.

Remove from heat; skim off foam quickly. Pour jelly immediately into hot containers and seal (p. 4).

Makes 4 or 5 eight-ounce glasses.

Blackberry Jelly

with powdered pectin

3½ cups blackberry juice (about 3 quart boxes berries)
1 package powdered pectin
4½ cups sugar

- **To prepare juice.** Sort and wash fully ripe berries; remove any stems or caps. Crush berries and extract juice (p. 6).
- **To make jelly.** Measure juice into kettle. Add pectin and stir

well. Place on high heat and, stirring constantly, bring quickly to a full rolling boil that cannot be stirred down.

Add sugar, continue stirring, and heat again to a full rolling boil. Boil hard for 1 minute.

Remove from heat; skim off foam quickly. Pour jelly immediately into hot containers and seal (p. 4).

Makes 5 or 6 eight-ounce glasses.

Blackberry Jelly

with liquid pectin

4 cups blackberry juice (about 3 quart boxes berries)
7½ cups sugar
1 bottle liquid pectin

• **To prepare juice.** Sort and wash fully ripe berries; remove any stems or caps. Crush berries and extract juice (p. 6).

• **To make jelly.** Measure juice into a kettle. Stir in sugar. Place on high heat and, stirring constantly, bring quickly to a full rolling boil that cannot be stirred down.

Add pectin and heat again to a full rolling boil. Boil hard for 1 minute.

Remove from heat; skim off foam quickly. Pour jelly immediately into hot containers and seal (p. 4).

Makes 8 or 9 eight-ounce glasses.

Blackberry Jelly

without added pectin

8 cups blackberry juice (about 5 quart boxes and 1½ cups water)
6 cups sugar

• **To prepare juice.** Select about one-fourth underripe and three-fourths ripe berries. Sort and wash; remove any stems or caps. Crush berries, add water, cover and bring to boil on high heat. Reduce heat and simmer for 5 minutes. Extract juice (p. 6).

• **To make jelly.** Measure juice into a kettle. Add sugar and stir well. Boil over high heat to 8° F. above the boiling point of water, or until jelly mixture sheets from a spoon.

Remove from heat; skim off foam quickly. Pour jelly immediately into hot containers and seal (p. 4).

Makes 7 or 8 eight-ounce glasses.

Orange Jelly

made from frozen concentrated juice

3¼ cups sugar
1 cup water
3 tablespoons lemon juice
½ bottle liquid pectin
1 six-ounce can (¾ cup) frozen concentrated orange juice

Stir the sugar into the water. Place on high heat and, stirring constantly, bring quickly to a full rolling boil that cannot be stirred down. Add lemon juice. Boil hard for 1 minute.

Remove from heat. Stir in pectin. Add thawed concentrated orange juice and mix well.

Pour jelly immediately into hot containers and seal (p. 4).

Makes 4 or 5 eight-ounce glasses.

How to make jelly with liquid pectin

Strawberry Jelly

Select fully ripe sound strawberries. About 3 quart boxes are needed for each batch of jelly. Sort the berries. Wash about 1 quart at a time by placing berries in a wire basket and moving the basket up and down several times in cold water. Drain the berries.

(78430-B)

Remove caps and crush the berries. Place crushed berries, a small amount at a time, in a damp jelly bag or double thickness of cheesecloth held in a colander over a bowl.

(78431-B)

Bring the edges of the cloth together and twist tightly. Press or squeeze to extract the juice. Strain the juice again through two thicknesses of damp cheesecloth without squeezing.

(78432-B)

Measure 4 cups of juice into a large kettle. Add 7½ cups of sugar to the juice; stir to dissolve the sugar.

Place the kettle over high heat and, stirring constantly, bring the mixture quickly to a full rolling boil that cannot be stirred down.

(78433-B)

Add 1 bottle of liquid pectin. Again, bring to a full rolling boil and boil hard for 1 minute.

Remove from heat and skim off foam quickly. If allowed to stand, the jelly may start to "set" in the kettle.

(78434-B)

Pour jelly immediately into hot glasses to ½ inch of top. Cover each glass with a ⅛ inch layer of paraffin. Cool glasses on a metal rack or folded cloth, then cover with metal or paper lids, label, and store in a cool, dry place.

(78436-B)

How to make jelly without added pectin

Apple Jelly

Use tart, firm apples. It takes about 3 pounds for a batch of jelly; about one-fourth of them should be underripe. Sort and wash the apples. Remove stems and blossom ends and cut apples into small pieces. Do not pare or core.

(9953-D)

Put apples into a kettle. Add 1 cup water per pound of apples. Cover, bring to boil on high heat. Reduce heat and simmer until apples are tender, about 20 to 25 minutes, depending on the firmness or ripeness of the fruit.

(78437-B)

Put cooked apples into a jelly bag and allow to drip, or press to remove juice. Strain pressed juice through two thicknesses of damp cheesecloth without squeezing.

(9954-D)

Measure 4 cups of the apple juice into a large kettle. Add 3 cups of sugar and 2 tablespoons of lemon juice, if desired. Stir to dissolve the sugar.

(78438-B)

Place on high heat and boil rapidly to 8° F. above the boiling point of water, or until jelly mixture sheets from a spoon. Remove from heat. Skim off foam.

(78439-B)

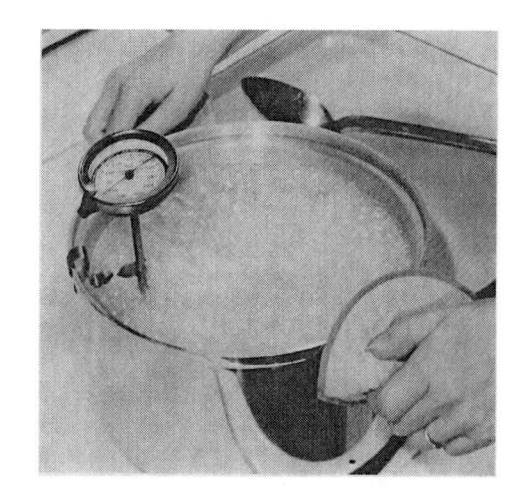

Pour jelly immediately into hot containers. Fill glasses (right) to ½ inch from top and cover with ⅛ inch layer of paraffin.

Or fill canning jars to ⅛ inch from top; wipe rims of jars. Cover with clean, hot metal lid with sealing compound next to glass. Screw metal band down tight.

Cool glasses or jars on a metal rack or folded cloth. Cover jelly glasses with metal or paper lids. Label and store in a cool, dry place.

(1174W1699-17)

How to Make Jellies, Jams, and Preserves at Home (13) 69

Cherry Jelly

with powdered pectin

3½ cups cherry juice (about 3 pounds or 2 quart boxes sour cherries and ½ cup water)
1 package powdered pectin
4½ cups sugar

- **To prepare juice.** Select fully ripe cherries. Sort, wash, and remove stems; do not pit. Crush cherries, add water, cover, bring to boil on high heat. Reduce heat and simmer for 10 minutes. Extract juice (p. 6).
- **To make jelly.** Measure juice into a kettle. Add pectin and stir well. Place on high heat and, stirring constantly, bring quickly to a full rolling boil that cannot be stirred down.

Add sugar, continue stirring, and heat again to a full rolling boil. Boil hard for 1 minute.

Remove from heat; skim off foam quickly. Pour jelly immediately into hot containers and seal (p. 4).

Makes about 6 eight-ounce glasses.

Cherry Jelly

with liquid pectin

3 cups cherry juice (about 3 pounds or 2 quart boxes sour cherries and ½ cup water)
7 cups sugar
1 bottle liquid pectin

- **To prepare juice.** Select fully ripe cherries. Sort, wash, and remove stems; do not pit. Crush cherries, add water, cover, bring to boil quickly. Reduce heat and simmer 10 minutes. Extract juice (p. 6).
- **To make jelly.** Measure juice into a kettle. Stir in sugar. Place on high heat and, stirring constantly, bring quickly to a full rolling boil that cannot be stirred down.

Add pectin; heat again to full rolling boil. Boil hard 1 minute.

Remove from heat; skim off foam quickly. Pour jelly immediately into hot containers and seal (p. 4).

Makes about 8 eight-ounce glasses.

Grape Jelly

with liquid pectin

4 cups grape juice (about 3½ pounds Concord grapes and ½ cup water)
7 cups sugar
½ bottle liquid pectin

- **To prepare juice.** Sort, wash, and remove stems from fully ripe grapes. Crush grapes, add water, cover, and bring to boil on high heat. Reduce heat and simmer for 10 minutes. Extract juice (p. 6).

To prevent formation of tartrate crystals in the jelly, let juice stand in a cool place overnight, then strain through two thicknesses of damp cheesecloth to remove crystals.

- **To make jelly.** Measure juice into a kettle. Stir in sugar. Place on high heat and, stirring constantly, bring quickly to a full rolling boil that cannot be stirred down.

Add pectin and heat again to a full rolling boil. Boil hard for 1 minute.

Remove from heat; skim off foam quickly. Pour jelly immediately into hot containers and seal (p. 4).

Makes 8 or 9 eight-ounce glasses.

Grape Jelly

made from frozen concentrated juice

6½ cups sugar
2½ cups water
1 bottle liquid pectin
3 six-ounce cans (2¼ cups) frozen concentrated grape juice

Stir sugar into water. Place on high heat and, stirring constantly, bring quickly to a full rolling boil that cannot be stirred down. Boil hard for 1 minute.

Remove from heat. Stir in pectin. Add thawed concentrated grape juice and mix well. Pour jelly immediately into hot containers and seal (p. 4).

Makes about 10 eight-ounce glasses.

Grape Jelly

with powdered pectin

5 cups grape juice (about 3½ pounds Concord grapes and 1 cup water)
1 package powdered pectin
7 cups sugar

- **To prepare juice.** Sort, wash, and remove stems from fully ripe grapes. Crush grapes, add water, cover, and bring to boil on high heat. Reduce heat and simmer for 10 minutes. Extract juice (p. 6). To prevent formation of tartrate crystals in the jelly, let juice stand in a cool place overnight, then strain through two thicknesses of damp cheesecloth to remove crystals that have formed.
- **To make jelly.** Measure juice into a kettle. Add pectin and stir well. Place on high heat and, stirring constantly, bring quickly to a full rolling boil that cannot be stirred down.

Add sugar, continue stirring, and bring again to a full rolling boil. Boil hard for 1 minute.

Remove from heat; skim off foam quickly. Pour jelly immediately into hot containers and seal (p. 4).

Makes 8 or 9 eight-ounce glasses.

Crabapple Jelly

without added pectin

4 cups crabapple juice (about 3 pounds crabapples and 3 cups water)
4 cups sugar

- **To prepare juice.** Select firm, crisp crabapples, about one-fourth underripe and three-fourths fully ripe. Sort, wash, and remove stem and blossom ends; do not pare or core. Cut crabapples into small pieces. Add water, cover, and bring to boil on high heat. Reduce heat and simmer for 20 to 25 minutes, or until crabapples are soft. Extract juice (p. 6).
- **To make jelly.** Measure juice into a kettle. Add sugar and stir well. Boil over high heat to 8° F. above the boiling point of water, or until mixture sheets from a spoon.

Remove from heat; skim off foam quickly. Pour jelly immediately into hot containers and seal (p. 4).

Makes 4 or 5 eight-ounce glasses.

Mint Jelly

with liquid pectin

1 cup chopped mint leaves and tender stems
1 cup water
½ cup cider vinegar
3½ cups sugar
5 drops green food coloring
½ bottle liquid pectin

- **To prepare mint.** Wash and chop mint. Pack solidly in a cup.
- **To make jelly.** Measure mint into a kettle. Add vinegar, water, and sugar; stir well. Place on high heat and, stirring constantly, bring quickly to a full rolling boil that cannot be stirred down.

Add food coloring and pectin; heat again to a full rolling boil. Boil hard for ½ minute.

Remove from heat. Skim. Strain through two thicknesses of damp cheesecloth. Pour jelly immediately into hot containers and seal (p. 4).

Makes 3 or 4 eight-ounce glasses.

Mixed Fruit Jelly

with liquid pectin

2 cups cranberry juice (about 1 pound cranberries and 2 cups water)
2 cups quince juice (about 2 pounds quince and 4 cups water)
1 cup apple juice (about ¾ pound apples and ¾ cup water)
7½ cups sugar
½ bottle liquid pectin

- **To prepare fruit.** Sort and wash fully ripe cranberries. Add water, cover, and bring to a boil on high heat. Reduce heat and simmer for 20 minutes. Extract juice (p. 6).

Sort and wash quince. Remove stem and blossom ends; do not pare or core. Slice very thin or cut into small pieces. Add water, cover, and bring to a boil on high heat. Reduce heat and simmer for 25 minutes. Extract juice (p. 6).

Sort and wash apples. Remove stem and blossom ends; do not pare or core. Cut into small pieces. Add water, cover, and bring to a boil on high heat. Reduce heat and simmer 20 minutes. Extract juice (p. 6).

NOTE: These juices may be prepared when the fruits are in season and then frozen or canned until the jelly is made.

- **To make jelly.** Measure juices into a kettle. Stir in sugar. Place on high heat and, stirring constantly, bring quickly to a full rolling boil that cannot be stirred down.

Add pectin and return to a full rolling boil. Boil hard for 1 minute.

Remove from heat; skim off foam quickly. Pour jelly immediately into hot containers and seal (p. 4).

Makes 9 or 10 eight-ounce glasses.

Plum Jelly

with liquid pectin

4 cups plum juice (about 4½ pounds plums and ½ cup water)
7½ cups sugar
½ bottle liquid pectin

● **To prepare juice.** Sort and wash fully ripe plums and cut in pieces; do not peel or pit. Crush fruit, add water, cover, and bring to boil over high heat. Reduce heat and simmer for 10 minutes. Extract juice (p. 6).

● **To make jelly.** Measure juice into a kettle. Stir in sugar. Place on high heat and, stirring constantly, bring quickly to a full rolling boil that cannot be stirred down.

Add pectin; bring again to full rolling boil. Boil hard 1 minute.

Remove from heat; skim off foam quickly. Pour jelly immediately into hot containers and seal (p. 4).

Makes 7 or 8 eight-ounce glasses.

Plum Jelly

with powdered pectin

5 cups plum juice (about 4½ pounds plums and 1 cup water)
1 package powdered pectin
7 cups sugar

● **To prepare juice.** Sort and wash fully ripe plums and cut in pieces; do not peel or pit. Crush fruit, add water, cover, and bring to boil on high heat. Reduce heat and simmer for 10 minutes. Extract juice (p. 6).

● **To make jelly.** Measure juice into a kettle. Add pectin and stir well. Place on high heat and, stirring constantly, bring quickly to a full rolling boil that cannot be stirred down.

Add sugar, continue stirring, and heat again to a full rolling boil. Boil hard for 1 minute.

Remove from heat; skim off foam quickly. Pour jelly immediately into hot containers and seal (p. 4).

Makes 7 or 8 eight-ounce glasses.

Quince Jelly

without added pectin

3¾ cups quince juice (about 3½ pounds quince and 7 cups water)
¼ cup lemon juice
3 cups sugar

● **To prepare juice.** Select about one-fourth underripe and three-fourths fully ripe quince. Sort, wash, and remove stems and blossom ends; do not pare or core. Slice quince very thin or cut into small pieces. Add water, cover, and bring to boil on high heat. Reduce heat and simmer for 25 minutes. Extract juice (p. 6).

● **To make jelly.** Measure quince juice into a kettle. Add lemon juice and sugar and stir well. Boil over high heat to 8° F. above the boiling point of water, or until jelly mixture sheets from a spoon.

Remove from heat; skim off foam quickly. Pour jelly immediately into hot containers and seal (p. 4).

Makes about 4 eight-ounce glasses.

Strawberry Jelly

with liquid pectin

Follow directions for blackberry jelly with liquid pectin, page 9. (See also pp. 6 and 4.)

Strawberry Jelly

with powdered pectin

Follow directions for blackberry jelly with powdered pectin, page 8.

Spiced Orange Jelly

with powdered pectin

2 cups orange juice (about 5 medium oranges)
⅓ cup lemon juice (about 2 medium lemons)
⅔ cup water
1 package powdered pectin
2 tablespoons orange peel, finely chopped
1 teaspoon whole allspice
½ teaspoon whole cloves
4 sticks cinnamon, 2 inches long
3½ cups sugar

Mix orange juice, lemon juice, and water in a large saucepan. Stir in pectin.

Place orange peel, allspice, cloves, and cinnamon sticks loosely in a clean white cloth; tie with a string and add to fruit mixture.

Place on high heat and, stirring constantly, bring quickly to a full rolling boil that cannot be stirred down.

Add sugar, continue stirring, and heat again to a full rolling boil. Boil hard for 1 minute.

Remove from heat. Remove spice bag and skim off foam quickly. Pour jelly immediately into hot containers and seal (p. 4).

Makes 4 eight-ounce glasses.

Jams

Jam is smooth and thick and has the natural color and flavor of the fruit from which it is made. It has a softer consistency than jelly.

On the following pages are directions for making jams from various fruits and combinations of fruits.

Because the products contain fruit pulp or pieces of fruit, they tend to stick to the kettle during cooking and require constant stirring to prevent scorching.

To help prevent fruit from floating in jam, remove cooked mixture from heat and stir gently at frequent intervals for 5 minutes.

With added pectin

For jams, as for jellies, the method of combining ingredients varies with the form of pectin used. Powdered pectin is mixed with the unheated crushed fruit; liquid pectin is added to the cooked fruit and sugar mixture immediately after it is removed from the heat.

Cooking time is the same for all the products—1 minute at a full boil. The full-boil stage is reached when bubbles form over the entire surface of the mixture.

With added pectin, jams can be made without cooking from some fresh or frozen fruits (see recipe, p. 26).

Without added pectin

Jams made without added pectin require longer cooking than those with added pectin. The most reliable way to judge doneness is to use a thermometer. Before making the product, take the temperature of boiling water. Cook the mixture to a temperature 9° F. higher than the boiling point of water. It is important to stir the mixture thoroughly just before taking the temperature, to place the thermometer vertically at the center of kettle, and to have the bulb covered with fruit mixture but not touching the bottom of the kettle. Read the thermometer at eye level.

If you have no thermometer, cook products made without added pectin until they have thickened somewhat. In judging thickness allow for the additional thickening of the mixture as it cools. The refrigerator test suggested for jelly may be used (see p. 8).

Blackberry Jam

with powdered pectin

6 cups crushed blackberries (about 3 quart boxes berries)
1 package powdered pectin
8½ cups sugar

- **To prepare fruit.** Sort and wash fully ripe berries; remove any stems or caps. Crush berries. If they are very seedy, put part or all of them through a sieve or food mill.
- **To make jam.** Measure crushed berries into a kettle. Add pectin and stir well. Place on high heat and, stirring constantly, bring quickly to a full boil with bubbles over the entire surface.

Add sugar, continue stirring, and heat again to a full bubbling boil. Boil hard for 1 minute, stirring constantly. Remove from heat; skim.

Fill and seal containers (p. 4).

Process 5 minutes in boiling water bath (p. 5).

Makes 11 or 12 half-pint jars.

Blackberry Jam

with liquid pectin

Follow directions for strawberry jam with liquid pectin, page 25. Put very seedy blackberries through a sieve or food mill.

How to make jam with powdered pectin

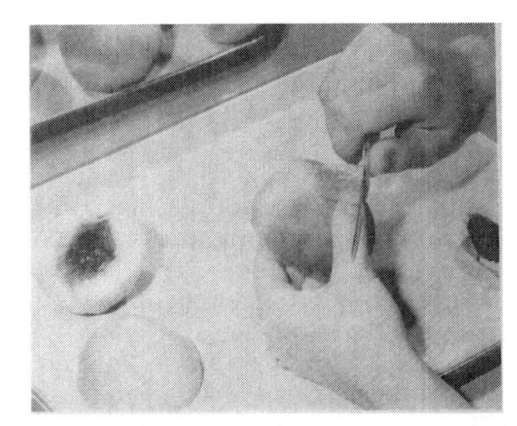

Peach Jam

Sort and wash fully ripe peaches. Remove stems, skins, and pits.

(78442-B)

Crush or chop the peaches. A stainless steel potato masher is useful for this purpose.

(78443-B)

Measure 3¾ cups of crushed peaches into a large kettle.

(78444-B)

76 (20) How to Make Jellies, Jams, and Preserves at Home

Add one package of powdered pectin and ¼ cup of lemon juice. Stir well to dissolve the pectin. Place on high heat and, stirring constantly, bring quickly to a full boil with bubbles over the entire surface.

(9955-D)

Stir in 5 cups of sugar, continue stirring, and heat again to a full bubbling boil. Boil hard for 1 minute, stirring constantly to prevent sticking.

Remove jam from heat and skim and stir alternately for 5 minutes to help prevent fruit from floating.

(78445-B)

Pour the jam into hot canning jars to ⅛ inch from top. Place clean, hot metal lid on, with sealing compound next to glass. Screw metal band down tight. Process 5 minutes in boiling water bath (p. 5). Cool jars on a metal rack or folded cloth, then label and store in a cool, dry place.

(1174W1700-17)

How to Make Jellies, Jams, and Preserves at Home (21) 77

Cherry Jam

with liquid pectin

4½ cups ground or finely chopped pitted cherries (about 3 pounds or 2 quart boxes sour cherries)
7 cups sugar
1 bottle liquid pectin

- **To prepare fruit.** Sort and wash fully ripe cherries; remove stems and pits. Grind cherries or chop fine.
- **To make jam.** Measure prepared cherries into a kettle. Add sugar and stir well. Place on high heat and, stirring constantly, bring quickly to a full boil with bubbles over the entire surface. Boil hard for 1 minute, stirring constantly.

Remove from heat and stir in the pectin. Skim.

Fill and seal containers (p. 4).

Process 5 minutes in boiling water bath (p. 5).

Makes about 8 half-pint jars.

Cherry Jam

with powdered pectin

4 cups ground or finely chopped pitted cherries (about 3 pounds or 2 quart boxes sour cherries)
1 package powdered pectin
5 cups sugar

- **To prepare fruit.** Sort and wash fully ripe cherries; remove stems and pits. Grind cherries or chop fine.
- **To make jam.** Measure prepared cherries into a kettle. Add pectin and stir well. Place on high heat and, stirring constantly, bring quickly to a full boil with bubbles over the entire surface.

Add sugar, continue stirring, and heat again to a full bubbling boil. Boil hard for 1 minute, stirring constantly. Remove from heat; skim.

Fill and seal containers (p. 4).

Process 5 minutes in boiling water bath (p. 5).

Makes 6 half-pint jars.

Mint–Pineapple Jam

with liquid pectin

1 20-oz. can crushed pineapple
¾ cup water
¼ cup lemon juice
7½ cups sugar
1 bottle liquid pectin
½ teaspoon mint extract
Few drops green coloring

Place crushed pineapple in a kettle. Add water, lemon juice, and sugar and stir well. Place on high heat and, stirring constantly, bring quickly to a full boil with bubbles over the entire surface. Boil hard for 1 minute, stirring constantly. Remove from heat; add pectin, flavor extract, and coloring. Skim.

Fill and seal containers (p. 4).

Process 5 minutes in boiling water bath (p. 5).

Makes 9 or 10 half-pint jars.

Variation. Use 10 drops oil of spearmint instead of extract.

Ginger-Peach Jam

Use either of the peach jam recipes (p. 23). Add 1 to 2 ounces of finely chopped candied ginger, as desired, to crushed peaches before adding pectin.

Fig Jam

with liquid pectin

4 cups crushed figs (about 3 pounds figs)
½ cup lemon juice
7½ cups sugar
½ bottle liquid pectin

- **To prepare fruit.** Sort and wash fully ripe figs; remove stem ends. Crush or grind fruit.
- **To make jam.** Place crushed figs into a kettle. Add sugar and stir well. Place on high heat and, stirring constantly, bring quickly to a full boil with bubbles over the entire surface. Boil hard for 1 minute, stirring constantly.

Remove from heat. Stir in pectin. Skim.

Fill and seal containers (p. 4).

Process 5 minutes in boiling water bath (p. 5).

Makes about 9 half-pint jars.

Peach Jam

with liquid pectin

4¼ cups crushed peaches (about 3½ pounds peaches)
¼ cup lemon juice
7 cups sugar
½ bottle liquid pectin

- **To prepare fruit.** Sort and wash fully ripe peaches. Remove stems, skins, and pits. Crush peaches.
- **To make jam.** Measure crushed peaches into a kettle. Add lemon juice and sugar and stir well. Place on high heat and, stirring constantly, bring quickly to a full boil with bubbles over the entire surface. Boil hard for 1 minute, stirring constantly.

Remove from heat; stir in pectin. Skim.

Fill and seal containers (p. 4).

Process 5 minutes in boiling water bath (p. 5).

Makes about 8 half-pint jars.

Peach Jam

with powdered pectin

3¾ cups crushed peaches (about 3 pounds peaches)
¼ cup lemon juice
1 package powdered pectin
5 cups sugar

- **To prepare fruit.** Sort and wash fully ripe peaches. Remove stems, skins, and pits. Crush peaches.
- **To make jam.** Measure crushed peaches into a kettle. Add lemon juice and pectin; stir well. Place on high heat and, stirring constantly, bring quickly to a full boil with bubbles over the entire surface.

Add sugar, continue stirring, and heat again to a full bubbling boil. Boil hard for 1 minute, stirring constantly. Remove from heat; skim.

Fill and seal containers (p. 4).

Process 5 minutes in boiling water bath (p. 5).

Makes about 6 half-pint jars.

Pineapple Jam

with liquid pectin

1 20-ounce can crushed pineapple
3 tablespoons lemon juice
3¼ cups sugar
½ bottle liquid pectin

Combine pineapple and lemon juice in a kettle.

Add sugar and stir well.

Place on high heat and, stirring constantly, bring quickly to a full boil with bubbles over the entire surface. Boil hard for 1 minute, stirring constantly.

Remove from heat; stir in pectin. Skim.

Let stand for 5 minutes.

Fill and seal containers (p. 4).

Process 5 minutes in boiling water bath (p. 5).

Makes 4 or 5 half-pint jars.

Rhubarb-Strawberry Jam

with liquid pectin

1 cup cooked red-stalked rhubarb (about 1 pound rhubarb and ¼ cup water)
2½ cups crushed strawberries (about 1½ quart boxes)
6½ cups sugar
½ bottle liquid pectin

- **To prepare fruit.** Wash rhubarb and slice thin or chop; do not peel. Add water, cover, and simmer until rhubarb is tender (about 1 minute).

Sort and wash fully ripe strawberries; remove stems and caps. Crush berries.

- **To make jam.** Measure prepared rhubarb and strawberries into a kettle. Add sugar and stir well. Place on high heat and, stirring constantly, bring quickly to a full boil with bubbles over the entire surface. Boil hard for 1 minute, stirring constantly.

Remove from heat and stir in pectin. Skim.

Fill and seal containers (p. 4).

Process 5 minutes in boiling water bath (p. 5).

Makes 7 or 8 half-pint jars.

Plum Jam

with liquid pectin

4½ cups crushed plums (about 2½ pounds plums)
7½ cups sugar
½ bottle liquid pectin

- **To prepare fruit.** Sort fully ripe plums, wash, cut into pieces, and remove pits. If flesh clings tightly to pits, cook plums slowly in a small amount of water for a few minutes until they are softened, then remove pits. Crush fruit.

- **To make jam.** Measure crushed plums into a kettle. Add sugar and stir well. Place on high heat and, stirring constantly, bring quickly to a full boil with bubbles over the entire surface. Boil hard for 1 minute, stirring constantly.

Remove from heat and stir in pectin. Skim.

Fill and seal containers (p. 4).

Process 5 minutes in boiling water bath (p. 5).

Makes about 8 half-pint jars.

Plum Jam

with powdered pectin

6 cups crushed plums (about 3½ pounds plums)
1 package powdered pectin
8 cups sugar

- **To prepare fruit.** Sort fully ripe plums, wash, cut into pieces, and remove pits. If flesh clings tightly to pits, cook plums slowly in a small amount of water for a few minutes until they are softened, then remove pits. Crush fruit.

- **To make jam.** Measure crushed

plums into a kettle. Add pectin and stir well. Place on high heat and, stirring constantly, bring quickly to a full boil with bubbles over the entire surface.

Add sugar, continue stirring, and heat again to a full bubbling boil. Boil hard for 1 minute. Remove from heat; skim.

Fill and seal containers (p. 4).

Process 5 minutes in boiling water bath (p. 5).

Makes about 9 half-pint jars.

Spiced Blueberry–Peach Jam

without added pectin

4 cups chopped or ground peaches (about 4 pounds peaches)
4 cups blueberries (about 1 quart fresh blueberries or 2 ten-ounce packages of unsweetened frozen blueberries)
2 tablespoons lemon juice
½ cup water
5½ cups sugar
½ teaspoon salt
1 stick cinnamon
½ teaspoon whole cloves
¼ teaspoon whole allspice

● **To prepare fruit.** Sort and wash fully ripe peaches; peel and remove pits. Chop or grind peaches.

Sort, wash, and remove any stems from fresh blueberries. Thaw frozen berries.

● **To make jam.** Measure fruits into a kettle; add lemon juice and water. Cover, bring to a boil, and simmer for 10 minutes, stirring occasionally.

Add sugar and salt; stir well. Add spices tied in cheesecloth. Boil rapidly, stirring constantly, to 9° F. above the boiling point of water, or until the mixture thickens.

Remove from heat; take out spices. Skim.

Fill and seal containers (p. 4).

Process 5 minutes in boiling water bath (p. 5).

Makes 6 or 7 half-pint jars.

Strawberry Jam

with liquid pectin

4 cups crushed strawberries (about 2 quart boxes strawberries)
7 cups sugar
½ bottle liquid pectin

● **To prepare fruit.** Sort and wash fully ripe strawberries; remove stems and caps. Crush berries.

● **To make jam.** Measure crushed strawberries into a kettle. Add sugar and stir well. Place on high heat and, stirring constantly, bring quickly to a full boil with bubbles over the entire surface. Boil hard for 1 minute, stirring constantly.

Remove from heat and stir in pectin. Skim.

Fill and seal containers (p. 4).

Process 5 minutes in boiling water bath (p. 5).

Makes 8 or 9 half-pint jars.

Strawberry Jam

with powdered pectin

5½ cups crushed strawberries (about 3 quart boxes strawberries)
1 package powdered pectin
8 cups sugar

● **To prepare fruit.** Sort and wash fully ripe strawberries; remove stems and caps. Crush berries.

● **To make jam.** Measure crushed strawberries into a kettle. Add pectin and stir well. Place on high heat and, stirring constantly, bring quickly to a full boil with bubbles over the entire surface.

Add sugar, continue stirring, and heat again to a full bubbling boil. Boil hard for 1 minute, stirring constantly. Remove from heat; skim.

Fill and seal containers (p. 4).

Process 5 minutes in boiling water bath (p. 5).

Makes 9 or 10 half-pint jars.

Tutti-Frutti Jam

with powdered pectin

3 cups chopped or ground pears (about 2 pounds pears)
1 large orange
¾ cup drained crushed pineapple
¼ cup chopped maraschino cherries (3-ounce bottle)
¼ cup lemon juice
1 package powdered pectin
5 cups sugar

- **To prepare fruit.** Sort and wash ripe pears; pare and core. Chop or grind the pears.

Peel orange, remove seeds, and chop or grind pulp.

- **To make jam.** Measure chopped pears into a kettle. Add orange, pineapple, cherries, and lemon juice. Stir in pectin. Place on high heat and, stirring constantly, bring quickly to a full boil with bubbles over the entire surface.

Add sugar, continue stirring, and heat again to a full bubbling boil. Boil hard for 1 minute, stirring constantly. Remove from heat; skim.

Fill and seal containers (p. 4).

Process 5 minutes in boiling water bath (p. 5).

Makes 6 or 7 half-pint jars.

Uncooked Berry Jam

with powdered pectin

2 cups crushed strawberries or blackberries (about 1 quart berries)
4 cups sugar
1 package powdered pectin
1 cup water

- **To prepare fruit.** Sort and wash fully ripe berries. Drain. Remove caps and stems; crush berries.
- **To make jam.** Place prepared berries in a large mixing bowl. Add sugar, mix well, and let stand for 20 minutes, stirring occasionally.

Dissolve pectin in water and boil for 1 minute. Add pectin solution to berry-and-sugar mixture; stir for 2 minutes.

Pour jam into freezer containers or canning jars, leaving ½ inch space at the top. Cover containers and let stand at room temperature for 24 hours or until jam has set.

Makes 5 or 6 half-pint jars.

- **To store.** Store uncooked jams in refrigerator or freezer. They can be held up to 3 weeks in a refrigerator or up to a year in a freezer. If kept at room temperature they will mold or ferment in a short time. Once a container is opened, jam should be stored in the refrigerator and used within a few days.

NOTE: If jam is too firm, stir to soften. If it tends to separate, stir to blend. If it is too soft, bring it to a boil. It will thicken on cooling.

Conserves, marmalades, preserves

Conserves are jam-like mixtures of two or more fruits plus nuts or raisins or both. They are rich in flavor and have a thick, but not sticky or gummy, consistency.

Marmalade is a mixture of fruits, usually including citrus, suspended in a clear, translucent jelly. The fruit is cut in small pieces or slices.

Preserves contain large or whole pieces of fruit saturated by a clear sirup of medium to thick consistency. The tender fruit retains its original size, shape, flavor, and color.

On the following pages are directions for making conserves, marmalades, and preserves from various fruits and combinations of fruits.

Because the products contain fruit pulp or pieces of fruit, they tend to stick to the kettle during cooking and require constant stirring to prevent scorching.

With added pectin

When powdered pectin is used in making conserves and marmalades, combine powdered pectin with unheated crushed fruit. Mix well. Bring to a full boil with bubbles over the entire surface. Add sugar and boil hard for 1 minute.

Without added pectin

Conserves, marmalades, and preserves made without added pectin require longer cooking than those with added pectin. The most reliable way to judge doneness is to use a thermometer. Before making the product, take the temperature of boiling water. Cook the mixture to a temperature 9° F. higher than the boiling point of water. It is important to stir the mixture thoroughly just before taking the temperature, to place thermometer vertically at the center of kettle, to have bulb covered with fruit mixture but not touching the bottom of the kettle. Read the thermometer at eye level.

If you have no thermometer, cook products made without added pectin until they have thickened somewhat. In judging thickness allow for the additional thickening of the mixture as it cools. The refrigerator test suggested for jelly may be used (see p. 8).

Apple Conserve

with powdered pectin

4½ cups finely chopped red apples (about 3 pounds apples)
½ cup water
¼ cup lemon juice
½ cup raisins
1 package powdered pectin
5½ cups sugar
½ cup chopped nuts

● **To prepare fruit.** Select tart apples. Sort and wash apples. Remove stem and blossom ends and core; do not pare. Chop apples fine.

● **To make conserve.** Combine apples, water, lemon juice, and raisins in a kettle. Add pectin and stir well. Place on high heat and, stirring constantly, bring quickly to a full boil with bubbles over the entire surface.

Add sugar, continue stirring, and heat again to a full bubbling boil. Boil hard for 1 minute, stirring constantly. Add nuts.

Remove from heat. If desired, add 3 or 4 drops of red food coloring. Skim.

Fill and seal containers (p. 4).

Process 5 minutes in boiling water bath (p. 5).

Makes 6 or 7 half-pint jars.

Apple Marmalade

without added pectin

8 cups thinly sliced apples (about 3 pounds)
1 orange
1½ cups water
5 cups sugar
2 tablespoons lemon juice

● **To prepare fruit.** Select tart apples. Wash, pare, quarter, and core the apples. Slice thin.

Quarter the orange, remove any seeds, and slice very thin.

● **To make marmalade.** Heat water and sugar until sugar is dissolved. Add the lemon juice and fruit. Boil rapidly, stirring constantly, to 9° F. above the boiling point of water, or until the mixture thickens. Remove from heat; skim.

Fill and seal containers (p. 4).

Process 5 minutes in boiling water bath (p. 5).

Makes 6 or 7 half-pint jars.

Damson Plum–Orange Conserve

with powdered pectin

3½ cups finely chopped damson plums (about 1½ pounds plums)
1 cup finely chopped oranges (1 or 2 oranges)
Peel of ½ orange
2 cups water
½ cup seedless raisins
1 package powdered pectin
7 cups sugar
½ cup chopped nuts

● **To prepare fruit.** Sort and wash plums; remove pits. Chop plums fine.

Peel and chop oranges. Shred peel of ½ orange very fine. Combine orange and peel, add the water, cover, and simmer for 20 minutes.

● **To make conserve.** Measure chopped plums into a kettle. Add orange, raisins, and pectin and stir well. Place on high heat and, stirring constantly, bring quickly to a full boil with bubbles over the entire surface.

Add sugar, continue stirring, and heat again to a full bubbling boil. Boil hard for 1 minute, stirring constantly. Stir in nuts. Remove from heat; skim.

Fill and seal containers (p. 4).

Process 5 minutes in boiling water bath (p. 5).

Makes 8 or 9 half-pint jars.

Apricot–Orange Conserve

without added pectin

3½ cups chopped drained apricots (about 2 20-oz. cans of unpeeled apricots or 1 pound dried apricots)
1½ cups orange juice (3 or 4 medium-size oranges)
Peel of ½ orange, shredded very fine
2 tablespoons lemon juice
3¼ cups sugar
½ cup chopped nuts

• **To prepare dried apricots.** Cook apricots uncovered in 3 cups water until tender (about 20 minutes); drain and chop.

• **To make conserve.** Combine all ingredients except nuts. Cook to 9° F. above the boiling point of water, or until thick, stirring constantly. Add nuts; stir well. Remove from heat; skim.

Fill and seal containers (p. 4).

Process 5 minutes in boiling water bath (p. 5).

Makes about 5 half-pint jars.

Grape Conserve

without added pectin

4½ cups grapes with skins removed (about 4 pounds Concord grapes)
1 orange
4 cups sugar
1 cup seedless raisins
½ teaspoon salt
Skins from grapes
1 cup nuts, chopped fine

• **To prepare fruit.** Sort and wash grapes; remove from stems. Slip skins from grapes; save skins. Measure skinned grapes into a kettle and boil, stirring constantly, for about 10 minutes, or until seeds show. Press through a sieve to remove seeds.

Chop orange fine without peeling it.

• **To make conserve.** Add orange, sugar, raisins, and salt to sieved grapes. Boil rapidly, stirring constantly, until the mixture begins to thicken (about 10 minutes).

Add grape skins and boil, stirring constantly, to 9° F. above the boiling point of water (about 10 minutes). Do not overcook; the mixture will thicken more on cooling. Add nuts and stir well. Remove from heat; skim.

Fill and seal containers (p. 4).

Process 5 minutes in boiling water bath (p. 5).

Makes 8 or 9 half-pint jars.

Citrus Marmalade

without added pectin

¾ cup grapefruit peel (½ grapefruit)
¾ cup orange peel (1 orange)
⅓ cup lemon peel (1 lemon)
1 quart cold water
Pulp of 1 grapefruit
Pulp of 4 medium-size oranges
⅓ cup lemon juice
2 cups boiling water
3 cups sugar

• **To prepare fruit.** Wash and peel fruit. Cut peel into thin strips. Add cold water and simmer in a covered pan until tender (about 30 minutes). Drain.

Remove seeds and membrane from peeled fruit. Cut fruit into small pieces.

● **To make marmalade.** Add boiling water to peel and fruit. Add sugar and boil rapidly to 9° F. above the boiling point of water (about 20 minutes), stirring frequently. Remove from heat; skim.

Fill and seal containers (p. 4).

Process 5 minutes in boiling water bath (p. 5).

Makes 3 or 4 half-pint jars.

Cranberry Marmalade

with powdered pectin

2 oranges
1 lemon
3 cups water
1 pound cranberries (about 4 cups)
1 box powdered pectin
7 cups sugar

● **To prepare fruit.** Peel oranges and lemon; remove half of white part of rinds. Finely chop or grind the remaining rinds. Put in large saucepan.

Add water, bring to a boil. Cover and simmer 20 minutes, stirring occasionally.

Chop peeled fruit. Sort and wash fully ripe cranberries. Add fruit to rind; cover and cook slowly 10 minutes longer.

● **To make marmalade.** Measure 6 cups of fruit into a large kettle. Add water to make 6 cups if necessary. Add pectin and stir well. Place on high heat and, stirring constantly, bring quickly to a full boil with bubbles over the entire surface.

Add sugar, continue stirring, and heat again to a full rolling boil. Boil hard for 1 minute, stirring constantly. Remove from heat; skim.

Fill and seal containers (p. 4).

Process 5 minutes in boiling water bath (p. 5).

Makes 10 or 11 half-pint jars.

Peach-Orange Marmalade

without added pectin

5 cups finely chopped or ground peaches (about 4 pounds peaches)
1 cup finely chopped or ground oranges (about 2 medium-size oranges)
Peel of 1 orange, shredded very fine
2 tablespoons lemon juice
6 cups sugar

● **To prepare fruit.** Sort and wash fully ripe peaches. Remove stems, skins, and pits. Finely chop or grind the peaches.

Remove peel, white portion, and seeds from oranges. Finely chop or grind the pulp.

● **To make marmalade.** Measure the prepared fruit into a kettle. Add remaining ingredients and stir well. Boil rapidly, stirring constantly, to 9° F. above the boiling point of water, or until the mixture thickens. Remove from heat; skim.

Fill and seal containers (p. 4).

Process 5 minutes in boiling water bath (p. 5).

Makes 6 or 7 half-pint jars.

Tomato Marmalade

without added pectin

3 quarts ripe tomatoes (about 5½ pounds tomatoes)
3 oranges
2 lemons
4 sticks cinnamon
1 tablespoon whole cloves
6 cups sugar
1 teaspoon salt

● **To prepare fruit.** Peel tomatoes; cut in small pieces. Drain. Slice oranges and lemons very thin; quarter the slices. Tie cinnamon and cloves in a cheesecloth bag.

- **To make marmalade.** Place tomatoes in a large kettle. Add sugar and salt; stir until dissolved. Add oranges, lemons, and spice bag. Boil rapidly, stirring constantly, until thick and clear (about 50 minutes). Remove from heat; skim.

Fill and seal containers (p. 4).

Process 5 minutes in boiling water bath (p. 5).

Makes about 9 half-pint jars.

Strawberry Preserves

6 cups prepared strawberries (about 2 quart boxes strawberries)
4½ cups sugar

- **To prepare fruit.** Select large, firm, tart strawberries. Wash and drain berries; remove caps.
- **To make preserves.** Combine prepared fruit and sugar in alternate layers and let stand for 8 to 10 hours or overnight in the refrigerator or other cool place.

Heat the fruit mixture to boiling, stirring gently. Boil rapidly, stirring as needed to prevent sticking.

Cook to 9° F. above the boiling point of water, or until the sirup is somewhat thick (about 15 or 20 minutes). Remove from heat; skim.

Fill and seal containers (p. 4).

Process 5 minutes in boiling water bath (p. 5).

Makes about 4 half-pint jars.

Questions and answers

High quality in jellied fruit products depends on so many complex factors that it is seldom possible to give just one answer to questions about problems in making these products. Using recipes from a reliable source—and following directions accurately—is the surest aid to success but does not guarantee it; it is impossible to assure uniform results because fruit varies widely in jellying quality.

The answers given here to questions commonly asked by homemakers who have had unsatisfactory results in making jellies and jams suggest possible reasons for lack of success.

Q. What makes jelly cloudy?

A. One or more of the following may cause cloudy jelly: Pouring jelly mixture into glasses too slowly. Allowing jelly mixture to stand before it is poured. Juice was not properly strained and so contained pulp. Jelly set too fast—usually the result of using too-green fruit.

Q. Why do crystals form in jelly?

A. Crystals throughout the jelly may be caused by too much sugar in the jelly mixture, or cooking the mixture too little, too slowly, or too long. Crystals that form at the top of jelly that has been opened and allowed to stand are caused by evaporation of liquid. Crystals in grape jelly may be tartrate crystals (see recipe for grape jelly, p. 14).

Q. What causes jelly to be too soft?

A. One or more of the following may be the cause: Too much juice in the mixture. Too little sugar. Mixture not acid enough. Making too big a batch at one time.

Q. What can be done to make soft jellies firmer?

A. Soft jellies can sometimes be improved by recooking according to the directions given below. It is best to recook only 4 to 6 cups of jelly at one time.

To remake with powdered pectin. Measure the jelly to be recooked. For each quart of jelly measure ¼ cup sugar, ¼ cup water, and 4 teaspoons powdered pectin. Mix the pectin and water and bring to boiling, stirring constantly to prevent scorching. Add the jelly and sugar. Stir thoroughly. Bring to a full rolling boil over high heat, stirring constantly. Boil mixture hard for ½ minute. Remove jelly from the heat, skim, pour into hot containers, and seal.

To remake with liquid pectin. Measure the jelly to be recooked. For each quart of jelly measure ¾ cup sugar, 2 tablespoons lemon juice, and 2 tablespoons liquid pectin. Bring jelly to boiling over high heat. Quickly add the sugar, lemon juice, and pectin and bring to a full rolling boil; stir constantly. Boil mixture hard for 1 minute. Remove jelly from the heat, skim, pour into hot containers, and seal.

To remake without added pectin. Heat the jelly to boiling and boil for a few minutes. Use one of the tests described on pages 7 and 8 to determine just how long to cook it. Remove jelly from the heat, skim, pour into hot containers, and seal.

Q. What makes jelly sirupy?

A. Too little pectin, acid, or sugar. A great excess of sugar can also cause sirupy jelly.

Q. What causes weeping jelly?

A. Too much acid. Layer of paraffin too thick. Storage place was too warm or storage temperature fluctuated.

Q. What makes jelly too stiff?

A. Too much pectin (fruit was not ripe enough or too much added pectin was used). Overcooking.

Q. What makes jelly tough?

A. Mixture had to be cooked too long to reach jellying stage, a result of too little sugar.

Q. What makes jelly gummy?

A. Overcooking.

Q. What causes fermentation of jelly?

A. Too little sugar, or improper sealing.

Q. Why does mold form on jelly or jam?

A. Because an imperfect seal has made it possible for mold and air to get into the container.

Q. What causes jelly or jam to darken at the top of the container?

A. Storage in too warm a place. Or a faulty seal that allows air to leak in.

Q. What causes fading?

A. Too warm a storage place or too long storage. Red fruits such as strawberries and raspberries are especially likely to fade.

Q. Why does fruit float in jam?

A. Fruit was not fully ripe, was not thoroughly crushed or ground, was not cooked long enough, or was not properly packed in glasses or jars. To prevent floating fruit, follow directions on page 21. (If glasses are used, stir jam before packing; if canning jars are used, shake jars gently after packing.)

Index

	Page		Page
Alcohol test for pectin	7	storing	5
Apple		Jelmeter test for pectin	7
conserve	28	Marmalade	
jelly	8	apple	28
marmalade	28	citrus	29
Apricot-orange conserve	29	cranberry	30
Blackberry		peach-orange	30
jam	19	pointers on making	27
jam, uncooked	26	storing	5
jelly	8, 9	tomato	30
Blueberry-peach jam, spiced	25	Mint jelly	16
Canned fruit, use of	4	Mint-pineapple jam	22
Canning jars, to use	3	Mixed fruit jelly	16
Cherry		Mold, causes	33
jam	22	Orange	
jelly	14	apricot-orange conserve	29
Citrus marmalade	29	damson plum-orange conserve	28
Cloudy jelly, causes	31	jelly	9
Conserve		jelly, spiced	18
apple	28	peach-orange marmalade	30
apricot-orange	29	Paraffin, use in sealing containers	5
damson plum-orange	28	Peach	
grape	29	blueberry-peach jam, spiced	25
pointers on making	27	ginger-peach jam	22
storing	5	jam	23
Containers		peach-orange marmalade	30
filling and sealing	4	Pear (See tutti-frutti jam.)	
kinds to use	3	Pectin, tests for	7
Crabapple jelly	15	Pineapple	
Cranberry marmalade	30	jam	23
Crystals in jelly, causes	32	mint-pineapple jam	22
Damson plum-orange conserve	28	Plum	
Darkening, causes	33	damson plum-orange conserve	28
Dried fruit, use of	4	jam	24
Equipment	3	jelly	16, 17
Fading, causes	33	Preserves	
Fermentation, causes	33	pointers on making	27
Fig jam	23	storing	5
Filling containers	4	strawberry	31
Floating fruit, causes	33	Processing	5
Frozen fruit, use of	4	Quince jelly	17
Fruit juice, to extract	6	Rhubarb-strawberry jam	24
Ginger-peach jam	22	Sealing containers	4
Glasses, to use	3	Sirupy jelly, causes	32
Grape		Soft jelly, causes and remedies	32
conserve	29	Spiced blueberry-peach jam	25
jelly	14, 15	Spiced orange jelly	18
Gummy jelly	33	Stiff jelly, causes	33
Ingredients, essential	1	Storage	5
Jam		Strawberry	
(See also blackberry, blueberry-peach, cherry, fig, ginger-peach, mint-pineapple, peach, pineapple, plum, rhubarb-strawberry, strawberry, tutti-frutti.)		jam	25
		jam, uncooked	26
		jelly	17, 18
		preserves	31
		rhubarb-strawberry jam	24
pointers on making	18	Tomato marmalade	30
storing	5	Tough jelly, causes	33
uncooked	26	Tutti-frutti jam	26
Jelly		Uncooked jam	
(See also apple, blackberry, cherry, crabapple, grape, mint, mixed fruit, orange, plum, quince, strawberry.)		Blackberry	26
		Strawberry	26
pointers on making	6	Weeping jelly, causes	33

☆ U.S. GOVERNMENT PRINTING OFFICE: 1974 O—527-629

Making PICKLES and RELISHES At Home

Pickle products truly add spice to meals or snacks. The skillful blending of spices, sugar, and vinegar with fruits and vegetables gives crisp, firm texture and pungent, sweet-sour flavor.

Pickles and relishes contribute some nutritive value, contain little or no fat, and, except for the sweet type, are low in calories.

Although food markets today offer a wide variety of pickles and relishes, many homemakers like to make their own pickle products when garden vegetables and fresh fruits are in abundant supply.

This bulletin gives specific directions for selecting and preparing pickling ingredients and for processing pickles and relishes. Included are basic recipes for the old-time favorites, such as pickled peaches, piccalilli, and sauerkraut; and for the newer fresh-pack or quick-process dills, sweet gherkins, crosscut pickle slices, and dilled green beans. Spices in these basic recipes can be increased or decreased to please family tastes.

Common causes of poor-quality pickles and spoilage in sauerkraut are pointed out.

Classes and Characteristics

Pickle products are classified on the basis of ingredients used and the method of preparation. There are four general classes.

Brined Pickles

Brined pickles, also called fermented pickles, go through a curing process of about 3 weeks. Dilled cucumbers and sauerkraut belong in this group. Other vegetables, such as green tomatoes, may also be cured in the same way as cucumbers. Curing changes cucumber color from a bright green to an olive or yellow green. The white interior of the fresh cucumber becomes uniformly translucent. A desirable flavor is developed during curing without being excessively sour, salty, or spicy. Cucumber dills may be flavored with garlic, if desired. The skin of the pickle is tender and firm, but not hard, rubbery, or shriveled. The inside is tender and firm, not soft or mushy.

Good sauerkraut (brined cabbage) has a pleasant tart and tangy flavor, and is free from any off-flavors or off-odors. It is crisp and firm in texture and has a bright, creamy-white color. The shreds are uniformly cut (about the thinness of a dime) and are free from large, coarse pieces of leaves or core.

Fresh-pack Pickles

Fresh-pack or quick-process pickles, such as crosscut cucumber slices and whole cucumber dills, sweet gherkins, and dilled green beans, are brined for several hours or overnight, then drained and combined with boiling-hot vinegar, spices, and other seasonings. These are quick and easy to prepare. They have a tart, pungent flavor. Seasonings can be selected to suit individual family preferences. Fresh-pack whole cucumbers are olive green, crisp, tender, and firm.

Fruit Pickles

Fruit pickles are usually prepared from whole fruits and simmered in a spicy, sweet-sour sirup. They should be bright in color, of uniform size, and tender and firm without being watery. Pears, peaches, and watermelon rind are prepared this way.

Relishes

Relishes are prepared from fruits and vegetables which are chopped, seasoned, and then cooked to desired consistency. Clear, bright color and uniformity in size of pieces make an attractive product. Relishes accent the flavor of other foods. They may be quite hot and spicy. Relishes include piccalilli, pepper-onion, tomato-apple chutney, tomato-pear chutney, horseradish, and corn relish.

A choice of pickles and relishes.

Ingredients for Successful Pickling

Satisfactory pickle products can be obtained only when good-quality ingredients are used and proper procedures are followed. Correct proportions of fruit or vegetable, sugar, salt, vinegar, and spices are essential. Alum and lime are not needed to make pickles crisp and firm if good-quality ingredients and up-to-date procedures are used.

Use tested recipes. Read the complete recipe before starting preparation. Make sure necessary ingredients are on hand. Measure or weigh all ingredients carefully.

Fruits and Vegetables

Selection.—Select tender vegetables and firm fruit. Pears and peaches may be slightly underripe for pickling. Use unwaxed cucumbers for pickling whole. The brine cannot penetrate waxed cucumbers. Sort for uniform size and select the size best suited for the recipe being followed.

Use fruits and vegetables as soon as possible after gathering from the orchard or garden, or after purchasing from the market. If the fruits and vegetables cannot be used immediately, refrigerate them, or spread them where they will be well ventilated and cool. This is particularly important for cucumbers because they deteriorate rapidly, especially at room temperatures.

Do not use fruits or vegetables that show even slight evidence of mold. Proper processing kills potential spoilage organisms, but does not destroy the off-flavor that may be produced by mold growth in the tissue.

Preparation.—Wash fruits and vegetables thoroughly in cold water, whether they are to be pared or left unpared. Use a brush and wash only a few at a time. Wash under running water or through several changes of water. Clinging soil may contain bacteria that are hard to destroy. Lift the fruits or vegetables out of the water each time, so soil that has been washed off will not be drained back over them. Rinse pan thoroughly between washings. Handle gently to avoid bruising.

Be sure to remove all blossoms from cucumbers. They may be a source of the enzymes responsible for softening of the cucumbers during fermentation.

Salt

Use pure granulated salt if available. Un-iodized table salt can be used, but the materials added to the salt to prevent caking may make the brine cloudy. Do not use iodized table salt; it may darken pickles.

Vinegar

Use a high-grade cider or white distilled vinegar of 4- to 6-percent acidity (40 to 60 grain). Vinegars of unknown acidity should not be used. Cider vinegar, with its mellow acid taste, gives a nice blending of flavors, but may darken white or light-colored fruits and vegetables. White distilled vinegar has a sharp, pungent, acetic acid taste and is desirable when light color is important, as with pickled pears, onions, and cauliflower.

Do not dilute the vinegar unless the recipe so specifies. If a less sour product is preferred, add sugar rather than decrease vinegar.

Sugar

Either white granulated or brown sugar may be used. White sugar gives a product with a lighter color, but brown sugar may be preferred for color.

Spices

The general term "spices" includes the sweet herbs and the pungent spices. Herbs are the leaves of aromatic plants grown in the Temperate Zone, and spices are the stems, leaves, roots, seeds, flowers, buds, and bark of aromatic plants grown in the tropics.

Use fresh spices for best flavor in pickles. Spices deteriorate and quickly lose their pungency in heat and humidity. If they cannot be used immediately, they should be stored in an airtight container in a cool place.

Equipment for Successful Pickling

Equipment of the right kind, size, and amount saves time and energy. Read the complete recipe before you start preparation and make sure you have the utensils and tools you need ready for use.

Utensils

For heating pickling liquids, use utensils of unchipped enamelware, stainless steel, aluminum, or glass. Do not use copper, brass, galvanized, or iron utensils; these metals may react with acids or salts and cause undesirable color changes in the pickles or form undesirable compounds.

For fermenting or brining, use a crock or stone jar, unchipped enamel-lined pan, or large glass jar, bowl, or casserole. Use a heavy plate or large glass lid, which fits inside the container, to cover vegetables in the brine. Use a weight to hold the cover down and keep vegetables below the surface of the brine. A glass jar filled with water makes a good weight.

Small utensils that add ease and convenience to home pickling include: Measuring spoons, large wood or stainless-steel spoons for stirring, measuring cups, sharp knives, large trays, tongs, vegetable peelers, ladle with lip for pouring, slotted spoon, footed colander or wire basket, large-mouthed funnel, food chopper or grinder, and wooden cutting board.

Water-Bath Canner

Any large metal container may be used for a water-bath canner if it—

• Is deep enough to allow for 1 or 2 inches of water above the tops of the jars, plus a little extra space for boiling.

• Has a close-fitting cover.

• Is equipped with a wire or wood rack with partitions to keep jars from touching each other and falling against the sides of the canner.

A steam-pressure canner can serve as a water bath. To use it for this purpose, set the cover in place without fastening it. Be sure the petcock is wide open so that steam escapes and pressure is not built up.

Glass Jars and Lids

Select jars and lids that are free of cracks, chips, rust, dents, or any defect that may prevent airtight seals and cause needless spoilage. Be sure to use the right kind of jar closure for the type of jar.

Do not use jars and lids from commercially canned foods. They are designed for use on special packing machines and are not suitable for home canning.

Wash glass jars in hot, soapy water. Rinse thoroughly with hot water.

Wash and rinse lids and bands. Metal lids with sealing compounds may need boiling or holding in boiling water for a few minutes until they are used. Follow the manufacturer's directions.

Use clean, new rubber rings of the right size for the jars. Do not test by stretching. Dip rubber rings in boiling water before putting them on the jars.

Scales

Household scales will be needed if the recipes specify ingredients by weight. They are necessary in making sauerkraut to insure correct proportions of salt and shredded cabbage.

A selection of jars and lids suitable for pickles and relishes is shown here. Jar types, from left to right, wide-mouth pint, regular pint, regular quart, 1½ pint with flared sides, wide-mouth quart, pint with flared sides, and ½ pint with flared sides. The closures, from left to right, flat metal lid with sealing compound, and metal screw band to fit regular jar; flat metal lid with sealing compound, and metal screw band to fit wide-mouth jar; and porcelain-lined zinc cap with shoulder rubber ring to fit regular jar.

Procedures for Successful Pickling

To insure acceptable quality and bacteriological safety of the finished pickle product, you must follow recommended procedures. Ingredients, time, and money may be wasted if you use outdated or careless canning procedures.

Filling Jars

Fill the jars firmly and uniformly with the pickle product. Avoid packing so tightly that the brine or sirup is prevented from filling around and over the product. Be sure to leave head space at the top of the jar, as recommended in recipe.

Wipe the rim and threads of the jar with a clean, hot cloth to remove any particles of food, seeds, or spices. Even a small particle may prevent an airtight seal.

When a porcelain-lined zinc cap with shoulder rubber ring is used, put the wet rubber ring on the jar shoulder before filling the jar. Do not stretch the rubber ring more than necessary. After filling the jar, wipe the rubber ring and jar rim and threads clean.

Closing Jars

The two-piece metal cap (flat metal lid and metal screw band) is the most commonly used closure. To use this type of closure, place the lid on the jar with the sealing compound next to the glass. Screw the metal band down tight by hand to hold the sealing compound against the glass. When band is screwed tight, this lid has enough "give" to let air escape during processing. Do not tighten screw band further after processing.

When using a porcelain-lined zinc cap with shoulder rubber ring, screw the cap down firmly against the wet rubber ring, then turn it back one-fourth inch. Immediately after processing and removal of the jar from the canner, screw the cap down tight to complete the seal.

If liquid has boiled out of a jar during processing, do not open it to add more liquid, because spoilage organisms may enter. Seal the jar just as it is.

Heat Treatment

Pickle products require heat treatment to destroy organisms that cause spoilage, and to inactivate enzymes that may affect flavor, color, and texture. Adequate heating is best achieved by processing the filled jars in a boiling-water bath.

Heat processing is recommended for all pickle products. There is always danger of spoilage organisms entering the food when it is transferred from kettle to jar. This is true even when the utmost caution is observed and is the reason open-kettle canning is not recommended.

Pack pickle products into glass jars according to directions given in the recipe. Adjust lids. Immerse the jars into actively boiling water in canner or deep kettle. Be sure the water comes an inch or two above the jars tops; add boiling water if necessary, but do not pour it directly on the jars. Cover the container with a close-fitting lid and bring the water back to boiling as quickly as possible. Start to count processing time when water returns to boiling, and continue to boil gently and steadily for the time recommended for the food being canned. Remove jars immediately and complete the seals if necessary. Set jars upright, several inches apart, on a wire rack to cool.

Processing procedures for fermented cucumbers and fresh-pack dills are slightly different from the usual water-bath procedures. For these products, start to count the processing time as soon as the filled jars are placed in the actively boiling water. This prevents development of a cooked flavor and a loss of crispness.

Processing times as given in the recipes are for altitudes less than 1,000 feet above sea level. At altitudes of 1,000 feet or above, you need to increase recommended processing times as follows:

Altitude (Feet)	Increase in processing time (minutes)
1,000	1
2,000	2
3,000	3
4,000	4
5,000	5
6,000	6
7,000	7
8,000	8
9,000	9
10,000	10

Cooling the Canned Pickles

Cool the jars top side up, on a wire rack, several inches apart to allow for free circulation of air. Keep the jars out of a draft. Do not cover.

Cool for 12 to 24 hours; remove metal screw bands carefully; then check jars for an airtight seal. If the center of the lid of the two-piece metal cap has a slight dip or stays down when pressed, the jar is sealed. Another test is to tap the center of the lid with a spoon. A clear, ringing sound means a good seal. A dull note, however, does not always mean a poor seal. Check for airtight seal by turning jar partly over. If there is no leakage, the jar may be stored.

If the porcelain-lined zinc cap with rubber ring has been used, check for airtight seal by turning the jar partly over. If there is no leakage, the seal is tight.

If a jar shows signs of leakage or a poor seal, use the product right away, or recan it. To recan, empty the jar, repack in another clean jar, and reprocess the product as before.

The metal screw bands from the two-piece metal caps may be used again. Remove them from the jars as soon as jars are cool. Sticking bands may be loosened by covering with a hot, damp cloth for a short time.

The metal lids from the two-piece metal caps may be used only one time.

Storing the Canned Pickles

Wipe the jars with a clean, damp cloth, and label with name of product and date.

Store the canned pickles in a dark,

dry, cool place where there is no danger of freezing. Freezing may crack the jars or break the seals, and let in bacteria that cause spoilage. Protect from light to prevent bleaching and possible deterioration of flavor.

Always be on the alert for signs of spoilage. Before opening a jar, examine it closely. A bulging lid or leakage may mean that the contents are spoiled.

When a jar is opened, look for other signs of spoilage, such as spurting liquid, mold, disagreeable odor, change in color, or an unusual softness, mushiness, or slipperiness of the pickle product. *If there is even the slightest indication of spoilage, do not eat or even taste the contents.* Dispose of the contents so that they cannot be eaten by humans or animals.

After emptying the jar of spoiled food, wash the jar in hot, soapy water and rinse. Boil in clean water for 15 minutes.

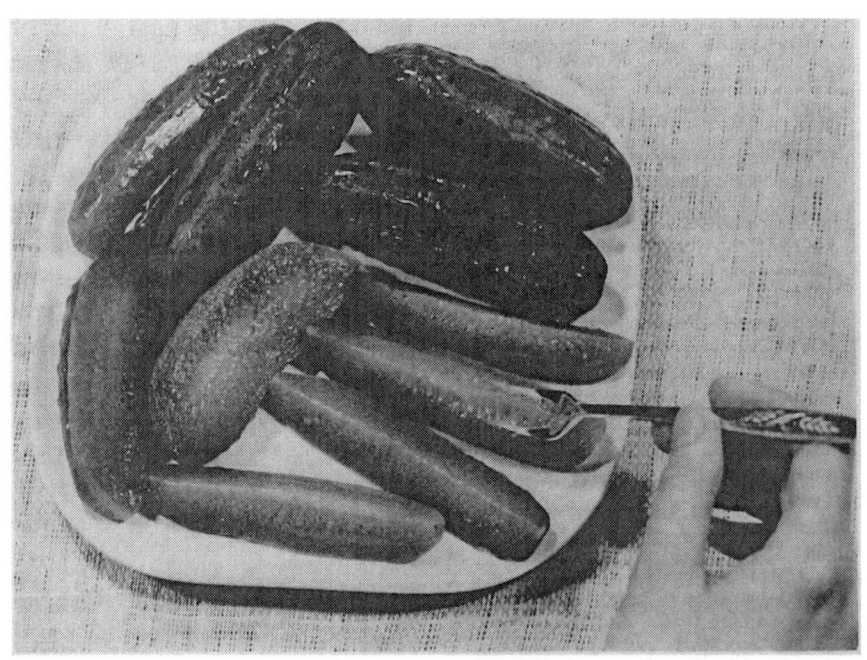

Heat-processed, brined dill pickles, ready for serving. Processing the pickles in a boiling-water bath destroys yeasts and molds in the jar and helps to preserve good texture and flavor in the pickles during storage for several months.

Recipes

VOLUME EQUIVALENTS

1 gallon = 4 quarts
1 quart = 4 cups
1 pint = 2 cups
1 cup = 16 tablespoons
1 tablespoon = 3 teaspoons

Brined dill pickles
Yield: 9 to 10 quarts

Cucumbers, 3 to 6 inches in length	20 pounds (about ½ bushel)
Whole mixed pickling spice	¾ cup
Dill plant, fresh or dried	2 to 3 bunches
Vinegar	2½ cups
Salt, pure granulated	1¾ cups
Water	2½ gallons

Cover cucumbers with cold water. Wash thoroughly, using a vegetable brush; handle gently to avoid bruising. Take care to remove any blossoms. Drain on rack or wipe dry.

Place half the pickle spices and a layer of dill in a 5-gallon crock or jar. Fill the crock with cucumbers to within 3 or 4 inches of the top. Place a layer of dill and remaining spices over the top of cucumbers. (Garlic may be added, if desired.) Thoroughly mix the vinegar, salt, and water and pour over the cucumbers.

Cover with a heavy china or glass plate or lid that fits inside the crock.

Use a weight to hold the plate down and keep the cucumbers under the brine. A glass jar filled with water makes a good weight. Cover loosely with clean cloth. Keep pickles at room temperature and remove scum daily when formed. Scum may start forming in 3 to 5 days. Do not stir pickles, but be sure they are completely covered with brine. If necessary, make additional brine, using original proportions specified in recipe.

In about 3 weeks the cucumbers will have become an olive-green color and should have a desirable flavor. Any white spots inside the fermented cucumbers will disappear in processing.

The original brine is usually cloudy as a result of yeast development during the fermentation period. If this cloudiness is objectionable, fresh brine may be used to cover the pickles when packing

(Continued on p. 12.)

How to make
Brined Dill Pickles

DN-2049
Wash cucumbers thoroughly with a brush. Use several changes of cold water. Take care to remove all blossoms. Drain on rack.

DN-2048
Place half of the spices and a layer of dill on the bottom of a 5-gallon jar or crock. Fill the cucumbers to 3 or 4 inches from top. Cover with remaining dill and add rest of spices. Mix salt, vinegar, and water and pour over cucumbers.

DN-2047
Use a heavy plate or glass lid which fits inside the container to cover cucumbers. Use a weight to hold the cover down and keep the cucumbers under the brine. A glass jar filled with water makes a good weight.

Bubbles and the formation of scum indicate active fermentation. Scum should be removed daily.

After 3 weeks of fermentation the dills are ready for processing. Cloudiness of the brine results from yeast development during fermentation. Strain the brine before using.

Pack pickles firmly into clean, hot quart jars. Do not wedge tightly. Add several pieces of the dill to each jar. Cover with boiling brine to $\frac{1}{2}$ inch of top of jar; adjust lids. Place jars in boiling water and process for 15 minutes. Start to count processing time as soon as hot jars are placed into the actively boiling water.

Remove jars from the canner and complete seals if necessary. Set jars upright, several inches apart, on a wire rack to cool. Cloudiness of brine is typical when the original fermentation brine is used as the covering liquid.

them into jars; in making fresh brine use ½ cup salt and 4 cups vinegar to 1 gallon of water. The fermentation brine is generally preferred for its added flavor and should be strained before heating to boiling.

Pack the pickles, along with some of the dill, into clean, hot quart jars; add garlic, if desired. Avoid too tight a pack. Cover with boiling brine to ½ inch of the top of the jar. Adjust jar lids.

Process in boiling water for 15 minutes [1] (start to count the processing time as soon as hot jars are placed into the actively boiling water).

Remove jars and complete seals if necessary. Set jars upright, several inches apart, on a wire rack to cool.

Fresh-pack dill pickles
Yield: 7 quarts

Cucumbers, 3 to 5 inches in length, packed 7 to 10 per quart jar	17 to 18 pounds
5-percent brine (¾ cup pure granulated salt per gallon of water)	About 2 gallons
Vinegar	6 cups (1½ quarts)
Salt, pure granulated	¾ cup
Sugar	¼ cup
Water	9 cups (2¼ quarts)
Whole mixed pickling spice	2 tablespoons
Whole mustard seed	2 teaspoons per quart jar
Garlic, if desired	1 or 2 cloves per quart jar
Dill plant, fresh or dried	3 heads per quart jar
Or	
Dill seed	1 tablespoon per quart jar

Wash cucumbers thoroughly; scrub with vegetable brush; drain. Cover with the 5-percent brine (¾ cup salt per gallon of water). Let set overnight; drain.

Combine vinegar, salt, sugar, water, and mixed pickling spices that are tied in a clean, thin, white cloth; heat to boiling. Pack cucumbers into clean, hot quart jars. Add mustard seed, dill plant or seed, and garlic to each jar; cover with boiling liquid to within ½ inch of top of jar. Adjust jar lids.

Process in boiling water for 20 minutes [1] (start to count the processing time as soon as hot jars are placed into the actively boiling water).

Remove jars and complete seals if necessary. Set jars upright, several inches apart, on a wire rack to cool.

---

[1] Processing time is given for altitudes less than 1,000 feet above sea level. At altitudes of 1,000 feet or above, see table on p. 7.

Sweet gherkins
Yield: 7 to 8 pints

Cucumbers, 1½ to 3 inches in length	5 quarts (about 7 pounds)
Salt, pure granulated	½ cup
Sugar	8 cups (2 quarts)
Vinegar	6 cups (1½ quarts)
Turmeric	¾ teaspoon
Celery seed	2 teaspoons
Whole mixed pickling spice	2 teaspoons
Stick cinnamon	8 1-inch pieces
Fennel (if desired)	½ teaspoon
Vanilla (if desired)	2 teaspoons

First day

Morning.—Wash cucumbers thoroughly; scrub with vegetable brush; stem ends may be left on if desired. Drain cucumbers; place in large container and cover with boiling water.

Afternoon (6 to 8 hours later).—Drain; cover with fresh, boiling water.

Second day

Morning.—Drain; cover with fresh, boiling water.

Afternoon.—Drain; add salt; cover with fresh, boiling water.

Third day

Morning.—Drain; prick cucumbers in several places with table fork. Make sirup of 3 cups of the sugar and 3 cups of the vinegar; add turmeric and spices. Heat to boiling and pour over cucumbers. (Cucumbers will be partially covered at this point.)

Afternoon.—Drain sirup into pan; add 2 cups of the sugar and 2 cups of the vinegar to sirup. Heat to boiling and pour over pickles.

Fourth day

Morning.—Drain sirup into pan; add 2 cups of the sugar and 1 cup of the vinegar to sirup. Heat to boiling and pour over pickles.

Afternoon.—Drain sirup into pan; add remaining 1 cup sugar and the vanilla to sirup; heat to boiling. Pack pickles into clean, hot pint jars and cover with boiling sirup to ½ inch of top of jar. Adjust jar lids.

Process for 5 minutes [1] in boiling water (start to count processing time as soon as water returns to boiling). Remove jars and complete seals if necessary. Set jars upright, several inches apart, on a wire rack to cool.

---

[1] Processing time is given for altitudes less than 1,000 feet above sea level. At altitudes of 1,000 feet or above, see table on p. 7.

How to make
Crosscut Pickle Slices

DN-2054
Wash cucumbers thoroughly. Slice unpeeled cucumbers into 1/8- to 1/4-inch crosswise slices. Wash and remove skins from onions; slice into 1/8-inch slices.

DN-2053
Combine cucumber and onion slices with peeled garlic cloves. Add salt and mix thoroughly. Cover with crushed ice or ice cubes. Allow to stand 3 hours. Drain thoroughly; remove garlic cloves.

DN-2052
Combine sugar, spices, and vinegar; heat to boiling. Add drained cucumber and onion slices and heat 5 minutes.

DN-2050
Pack loosely into clean, hot pint jars to ½ inch of top of jar; adjust lids. Process in boiling water for 5 minutes. Start to count processing time as soon as water in canner returns to boiling.

DN-2051
Remove jars and complete seals if necessary. Set jars upright, several inches apart, on a wire rack or folded towel to cool.

Making Pickles and Relishes at Home (15) 105

Crosscut pickle slices

Yield: 7 pints

Cucumbers, medium size (about 6 pounds), sliced	4 quarts
Onions (12 to 15 small white, about 1 pound), sliced	1½ cups
Garlic cloves	2 large
Salt	⅓ cup
Ice, crushed or cubes	2 quarts (2 trays)
Sugar	4½ cups
Turmeric	1½ teaspoons
Celery seed	1½ teaspoons
Mustard seed	2 tablespoons
Vinegar, white	3 cups

Wash cucumbers thoroughly, using a vegetable brush; drain on rack. Slice unpeeled cucumbers into ⅛-inch to ¼-inch slices; discard ends. Add onions and garlic.

Add salt and mix thoroughly; cover with crushed ice or ice cubes; let stand 3 hours. Drain thoroughly; remove garlic cloves.

Combine sugar, spices, and vinegar; heat just to boiling. Add drained cucumber and onion slices and heat 5 minutes.

Pack hot pickles loosely into clean, hot pint jars to ½ inch of top. Adjust jar lids.

Process in boiling water for 5 minutes [1] (start to count processing time as soon as water in canner returns to boiling). Remove jars and complete seals if necessary. Set jars upright to cool.

Note: Sugar may be reduced to 4 cups, if a less sweet pickle is desired.

Tomato-apple chutney

Yield: 7 pints

Tomatoes (about 6 pounds), pared, chopped	3 quarts
Apples (about 5 pounds), pared, chopped	3 quarts
Raisins, seedless, white	2 cups
Onions (4 to 5 medium), chopped	2 cups
Green peppers (2 medium), chopped	1 cup
Brown sugar	2 pounds
Vinegar, white	1 quart
Salt	4 teaspoons
Ground ginger	1 teaspoon
Whole mixed pickling spice	¼ cup

Combine all ingredients except the whole spices. Place spices loosely in a clean, white cloth; tie with a string, and add to tomato-

---

[1] Processing time is given for altitudes less than 1,000 feet above sea level. At altitudes of 1,000 feet or above, see table on p. 7.

apple mixture. Bring to a boil; cook slowly, stirring frequently, until mixture is thickened (about 1 hour). Remove spice bag.

Pack the boiling-hot chutney into clean, hot pint jars to ½ inch of the top of the jar. Adjust jar lids. Process in boiling water for 5 minutes [1] (start to count processing time when water in canner returns to boiling). Remove jars and complete seals if necessary. Set jars upright, several inches apart, on a wire rack to cool.

Tomato-pear chutney

Yield: 3 to 4 jars (½ pint each)

Tomatoes, quartered, fresh or canned	2½ cups
Pears, diced, fresh or canned	2½ cups
Raisins, seedless, white	½ cup
Green pepper (1 medium), chopped	½ cup
Onions (1 or 2 medium), chopped	½ cup
Sugar	1 cup
Vinegar, white	½ cup
Salt	1 teaspoon
Ground ginger	½ teaspoon
Mustard, powdered, dry	½ teaspoon
Cayenne pepper	⅛ teaspoon
Pimiento, canned, chopped	¼ cup

When fresh tomatoes and pears are used, remove skins; include sirup when using canned pears.

Combine all ingredients except pimiento. Bring to a boil; cook slowly until thickened (about 45 minutes), stirring occasionally. Add pimiento and boil 3 minutes longer.

Pack the boiling-hot chutney into clean, hot jars, filling to the top. Seal tightly. *Store in refrigerator.*

If extended storage without refrigeration is desired, this product should be processed in boiling water. Pack the boiling-hot chutney into clean, hot jars to ½ inch of top of jar. Adjust jar lids. Process in boiling water for 5 minutes [1] (start to count processing time when water in canner returns to boiling). Remove jars and complete seals if necessary. Set jars upright, several inches apart, on a wire rack to cool.

Note: If a less spicy chutney is preferred, the amount of cayenne pepper may be reduced or omitted.

---

[1] Processing time is given for altitudes less than 1,000 feet above sea level. At altitudes of 1,000 feet or above, see table on p. 7.

Dilled green beans
Yield: 7 pints

Green beans, whole	4 pounds (about 4 quarts)
Hot red pepper, crushed	¼ teaspoon per pint jar
Whole mustard seed	½ teaspoon per pint jar
Dill seed	½ teaspoon per pint jar
Garlic	1 clove per pint jar
Vinegar	5 cups (1¼ quarts)
Water	5 cups (1¼ quarts)
Salt	½ cup

Wash beans thoroughly; drain and cut into lengths to fill pint jars. Pack beans into clean, hot jars; add pepper, mustard seed, dill seed, and garlic.

Combine vinegar, water and salt; heat to boiling. Pour boiling liquid over beans, filling to ½ inch of top of jar. Adjust jar lids.

Process in boiling water for 5 minutes [1] (start to count processing time as soon as water in canner returns to boiling). Remove jars and complete seals if necessary. Set jars upright, several inches apart, on a wire rack to cool.

Pickled peaches
Yield: 7 quarts

Sugar	3 quarts
Vinegar	2 quarts
Stick cinnamon	7 2-inch pieces
Cloves, whole	2 tablespoons
Peaches, small or medium size	16 pounds (about 11 quarts)

Combine sugar, vinegar, stick cinnamon, and cloves. (Cloves may be put in a clean cloth, tied with a string, and removed after cooking, if not desired in packed product.) Bring to a boil and let simmer covered, about 30 minutes.

Wash peaches and remove skins; dipping the fruit in boiling water for 1 minute, then quickly in cold water makes peeling easier. To prevent pared peaches from darkening during preparation, immediately put them into cold water containing 2 tablespoons each of salt and vinegar per gallon. Drain just before using.

Add peaches to the boiling sirup, enough for 2 or 3 quarts at a time, and heat for about 5 minutes. Pack hot peaches into clean, hot jars. Continue heating in sirup and packing peaches into jars. Add 1 piece of stick cinnamon and 2 to 3 whole cloves (if desired) to each jar. Cover peaches with boiling sirup to

---
[1] Processing time is given for altitudes less than 1,000 feet above sea level. At altitudes of 1,000 feet or above, see table on p. 7.

½ inch of top of jar. Adjust jar lids.

Process in boiling water for 20 minutes [1] (start to count processing time after water in canner returns to boiling). Remove jars and complete seals if necessary. Set jars upright, several inches apart, on a wire rack to cool.

Pickled pears
Yield: 7 to 8 pints

Sugar	2 quarts
Vinegar, white	1 quart
Water	1 pint
Stick cinnamon	8 2-inch pieces
Cloves, whole	2 tablespoons
Allspice, whole	2 tablespoons
Seckel pears	8 pounds (4 or 5 quarts)

Combine sugar, vinegar, water, and stick cinnamon; add cloves and allspice that are tied in a clean, thin white cloth. Bring to a boil and simmer, covered, about 30 minutes.

Wash pears, remove skins, and all of blossom end; the stems may be left on if desired. To prevent peeled pears from darkening during preparation, immediately put them into cold water containing 2 tablespoons each of salt and vinegar per gallon. Drain just before using.

Add pears to the boiling sirup and continue simmering for 20 to 25 minutes. Pack hot pears into clean, hot pint jars; add one 2-inch piece cinnamon per jar and cover with boiling sirup to ½ inch of top of jar. Adjust jar lids.

Process in boiling water for 20 minutes [1] (start to count processing time as soon as water in canner returns to boiling). Remove jars and complete seals if necessary. Set jars upright, several inches apart, on a wire rack to cool.

Kieffer pears are also frequently used for making fruit pickles.

To pickle Kieffer pears: Use 12 pounds Kieffer pears and reduce vinegar to 3 cups in recipe above. Wash the pears, peel, cut in halves or quarters, remove hard centers and cores. Boil pears for 10 minutes in water to cover. Use 1 pint of this liquid in place of the pint of water in recipe above. Finish in the same way as Seckel pears. Makes about 8 pints.

---

[1] Processing time is given for altitudes less than 1,000 feet above sea level. At altitudes of 1,000 feet or above, see table on p. 7.

Watermelon pickles
Yield: 4 to 5 pints

Watermelon rind (about 6 pounds, unpared, or ½ large melon)	3 quarts
Salt	¾ cup
Water	3 quarts
Ice cubes	2 quarts (2 trays)
Sugar	9 cups (2¼ quarts)
Vinegar, white	3 cups
Water	3 cups
Whole cloves	1 tablespoon (about 48)
Stick cinnamon	6 1-inch pieces
Lemon, thinly sliced, with seeds removed	1

Pare rind and all pink edges from the watermelon. Cut into 1-inch squares or fancy shapes as desired. Cover with brine made by mixing the salt with 3 quarts cold water. Add ice cubes. Let stand 5 or 6 hours.

Drain; rinse in cold water. Cover with cold water and cook until fork tender, about 10 minutes (do not overcook). Drain.

Combine sugar, vinegar, water, and spices (tied in a clean, thin white cloth). Boil 5 minutes and pour over the watermelon with spices; add lemon slices. Let stand overnight.

Heat watermelon in sirup to boiling and cook until watermelon is translucent (about 10 minutes). Pack hot pickles loosely into clean, hot pint jars. To each jar add 1 piece of stick cinnamon from spice bag; cover with boiling sirup to ½ inch of top of jar. Adjust jar lids.

Process in boiling water for 5 minutes [1] (start to count processing time when water in canner returns to boiling). Remove jars and complete seals if necessary. Set jars upright, several inches apart, on a wire rack to cool.

The sugar may be reduced to 8 cups, if a less sweet pickle is desired.

Note: Red or green coloring may be added to the sirup, if desired. Keep watermelon rind in plastic bags in refrigerator until enough rinds for one recipe are collected.

---

[1] Processing time is given for altitudes less than 1,000 feet above sea level. At altitudes of 1,000 feet or above, see table on p. 7.

Sauerkraut
Yield: 16 to 18 quarts

Cabbage	About 50 pounds
Salt, pure granulated	1 pound (1 ½ cups)

Remove the outer leaves and any undesirable portions from firm, mature, heads of cabbage; wash and drain. Cut into halves or quarters; remove the core. Use a shredder or sharp knife to cut the cabbage into thin shreds about the thickness of a dime.

In a large container, thoroughly mix 3 tablespoons salt with 5 pounds shredded cabbage. Let the salted cabbage stand for several minutes to wilt slightly; this allows packing without excessive breaking or bruising of the shreds.

Pack the salted cabbage firmly and evenly into a large clean crock or jar. Using a wooden spoon or tamper or the hands, press down firmly until the juice comes to the surface. Repeat the shredding, salting, and packing of cabbage until the crock is filled to within 3 or 4 inches of the top.

Cover cabbage with a clean, thin, white cloth (such as muslin) and tuck the edges down against the inside of the container. Cover with a plate or round paraffined board that just fits inside the container so that the cabbage is not exposed to the air. Put a weight on top of the cover so the brine comes to the cover but not over it. A glass jar filled with water makes a good weight.

A newer method of covering cabbage during fermentation consists of placing a plastic bag filled with water on top of the fermenting cabbage. The water-filled bag seals the surface from exposure to air, and prevents the growth of film yeast or molds. It also serves as a weight. For extra protection, the bag with the water in it can be placed inside another plastic bag.

Any bag used should be of heavyweight, watertight plastic and intended for use with foods.

The amount of water in the plastic bag can be adjusted to give just enough pressure to keep the fermenting cabbage covered with brine. See illustration on page 25.

Formation of gas bubbles indicates fermentation is taking place. A room temperature of 68° to 72° F. is best for fermenting cabbage. Fermentation is usually completed in 5 to 6 weeks.

To store: Heat sauerkraut to simmering (185° to 210° F.). Do not boil. Pack hot sauerkraut into clean, hot jars and cover with hot juice to ½ inch of top of jar. Adjust jar lids. Process in boiling-water bath, 15 minutes for pints, and 20 minutes for quarts.[1] Start to count processing time as soon as hot jars are placed into the actively boiling water.

Remove jars and complete seals if necessary. Set jars upright, several inches apart, to cool.

---

[1] Processing time is given for altitudes less than 1,000 feet above sea level. At altitudes of 1,000 feet or above, see table on p. 7.

How to make Sauerkraut

DN-2042
Remove the outer leaves from firm, mature heads of cabbage; wash and drain. Remove core or cut it into thin shreds.

DN-2041
Shred cabbage and weigh 5 pounds. Accuracy in weighing is important to insure correct proportion of cabbage to salt.

DN-2040
Measure 3 tablespoons pure granulated salt and sprinkle over 5 pounds shredded cabbage.

DN-2061
Allow the salted cabbage to stand a few minutes to wilt slightly. Mix well, with clean hands or a spoon, to distribute salt uniformly.

DN-2039
Pack the salted cabbage into container. Press firmly with wooden spoon, tamper, or with hands until the juices drawn out will just cover the shredded cabbage.

DN-2060
Place a water-filled plastic bag on top of the cabbage. A water-filled plastic bag fits snugly against the cabbage and against the sides of the container and prevents exposure to air.

Making Pickles and Relishes at Home (23) 113

DN-2062

When fermentation is complete, remove from container and heat in kettle to simmering temperature. Pack hot sauerkraut into clean, hot jars; cover with hot juice, filling to ½ inch of top of jar. Adjust lids. Place jars in boiling-water bath and process 15 minutes for pints and 20 minutes for quarts. Start to count the processing time as soon as hot jars are placed into the actively boiling water.

Remove jars from the canner and complete seals if necessary. Set jars upright, several inches apart, on a wire rack to cool.

DN-2063

Relishes for which ingredients are available throughout the year can be made up in small quantities for use within a period of 3 or 4 weeks. For such products, the boiling water-bath process may be omitted *but they must be stored in the refrigerator.* Recipes for two relishes that can be made in this way appear on this page.

Horseradish relish

Grated horseradish	1 cup
Vinegar, white	½ cup
Salt	¼ teaspoon

Wash horseradish roots thoroughly and remove the brown, outer skin. (A vegetable peeler is useful in removal of outer skin.) The roots may be grated, or cut into small cubes and put through a food chopper or a blender.

Combine ingredients. Pack into clean jars. Seal tightly. *Store in refrigerator.*

Pepper-onion relish

Yield: 5 jars (½ pint each)

Onions (6 to 8 large), finely chopped	1 quart
Sweet red peppers (4 or 5 medium), finely chopped	1 pint
Green peppers (4 or 5 medium), finely chopped	1 pint
Sugar	1 cup
Vinegar	1 quart
Salt	4 teaspoons

Combine all ingredients and bring to a boil. Cook until slightly thickened (about 45 minutes), stirring occasionally. Pack the boiling-hot relish into clean, hot jars; fill to top of jar. Seal tightly. *Store in refrigerator.*

If extended storage without refrigeration is desired, this product should be processed in a boiling-water bath. Pack the boiling-hot relish into clean, hot jars to ½ inch of top of jar. Adjust jar lids. Process in boiling water for 5 minutes [1] (start to count processing time when water in canner returns to boiling).

Remove jars and complete seals if necessary. Set jars upright, several inches apart, on a wire rack to cool.

---

[1] Processing time is given for altitudes less than 1,000 feet above sea level. At altitudes of 1,000 feet or above, see table on p. 7.

Piccalilli
Yield: 4 pints

Green tomatoes (about 16 medium), chopped	1 quart
Sweet red peppers (2 to 3 medium), chopped	1 cup
Green peppers (2 to 3 medium), chopped	1 cup
Onions (2 to 3 large), chopped	1½ cups
Cabbage (about 2 pounds), chopped	5 cups (1¼ quarts)
Salt	⅓ cup
Vinegar	3 cups
Brown sugar	2 cups, firmly packed
Whole mixed pickling spice	2 tablespoons

Combine vegetables, mix with salt, let stand overnight. Drain and press in a clean, thin, white cloth to remove all liquid possible.

Combine vinegar and sugar. Place spices loosely in a clean cloth; tie with a string. Add to vinegar mixture. Bring to a boil.

Add vegetables, bring to a boil, and simmer about 30 minutes, or until there is just enough liquid to moisten vegetables. Remove spice bag. Pack hot relish into clean, hot pint jars. Fill jars to ½ inch of top. Adjust lids.

Process in boiling water for 5 minutes [1] (start to count processing time as soon as water in canner returns to boiling).

Remove jars and complete seals if necessary. Set jars upright on a wire rack to cool.

Corn relish
Yield: 7 pints

Corn, whole kernel	2 quarts
Use fresh (16 to 20 medium-size ears) or frozen (whole kernel, six 10-ounce packages)	
Sweet red peppers (4 to 5 medium) diced	1 pint
Green peppers (4 to 5 medium), diced	1 pint
Celery (1 large bunch), chopped	1 quart
Onions (8 to 10 small, ¾ pound) chopped or sliced	1 cup
Sugar	1½ cups
Vinegar	1 quart
Salt	2 tablespoons
Celery seed	2 teaspoons
Mustard, powdered dry	2 tablespoons
Turmeric	1 teaspoon

Fresh corn.—Remove husks and silks. Cook ears of corn in boiling water for 5 minutes; remove and

[1] Processing time is given for altitudes less than 1,000 feet above sea level. At altitudes of 1,000 feet or above, see table on p. 7.

plunge into cold water. Drain; cut corn from cob. Do not scrape cob.

Frozen corn.—Defrost overnight in refrigerator or for 2 to 3 hours at room temperature. Place containers in front of a fan to hasten defrosting.

Combine peppers, celery, onions, sugar, vinegar, salt, and celery seed. Cover pan until mixture starts to boil, then boil uncovered for 5 minutes, stirring occasionally. Mix dry mustard and turmeric and blend with liquid from boiling mixture; add, with corn, to boiling mixture. Return to boiling and cook for 5 minutes, stirring occasionally.

This relish may be thickened by adding ¼ cup flour blended with ½ cup water at the time the corn is added for cooking. Frequent stirring will be necessary to prevent sticking and scorching.

Pack loosely while boiling hot into clean, hot pint jars, filling to ½ inch of top. Adjust jar lids.

Process in boiling water for 15 minutes [1] (start to count processing time as soon as water in canner returns to boiling). Remove jars and complete seals if necessary. Set jars upright, several inches apart, on a wire rack to cool.

Common Causes of Poor-Quality Pickles

Shriveled Pickles

Shriveling may result from using too strong a vinegar, sugar, or salt solution at the start of the pickling process. In making the very sweet or very sour pickles, it is best to start with a dilute solution and increase gradually to the desired strength.

Overcooking or overprocessing may also cause shriveling.

Hollow Pickles

Hollowness in pickles usually results from—

- Poorly developed cucumbers.
- Holding cucumbers too long before pickling.
- Too rapid fermentation.
- Too strong or too weak a brine during fermentation.

Soft or Slippery Pickles

These generally result from microbial action which causes spoilage. Once a pickle becomes soft it cannot be made firm. Microbial activity may be caused by—

- Too little salt or acid.
- Cucumbers not covered with brine during fermentation.
- Scum scattered throughout the brine during fermentation period.
- Insufficient heat treatment.
- A seal that is not airtight.
- Moldy garlic or spices.

Blossoms, if not entirely removed from the cucumbers before fermentation, may contain fungi or yeasts

---

[1] Processing time is given for altitudes less than 1,000 feet above sea level. At altitudes of 1,000 feet or above, see table on p. 7.

responsible for enzymatic softening of pickles.

Dark Pickles

Darkness in pickles may be caused by—
- Use of ground spices.
- Too much spice.
- Iodized salt.
- Overcooking.
- Minerals in water, especially iron.
- Use of iron utensils.

Common Causes of Spoilage in Sauerkraut

Off-flavors and off-odors develop when there is spoilage in sauerkraut. Spoilage in sauerkraut is indicated by undesirable color, off-odors, and soft texture.

Soft Kraut

Softness in sauerkraut may result from—
- Insufficient salt.
- Too high temperatures during fermentation.
- Uneven distribution of salt.
- Air pockets caused by improper packing.

Pink Kraut

Pink color in kraut is caused by growth of certain types of yeast on the surface of the kraut. These may grow if there is too much salt, an uneven distribution of salt, or if the kraut is improperly covered or weighted during fermentation.

Rotted Kraut

This condition in kraut is usually found at the surface where the cabbage has not been covered sufficiently to exclude air during fermentation.

Dark Kraut

Darkness in kraut may be caused by—
- Unwashed and improperly trimmed cabbage.
- Insufficient juice to cover fermenting cabbage.
- Uneven distribution of salt.
- Exposure to air.
- High temperatures during fermentation, processing, and storage.
- Long storage period.

Index to Recipes

	Page
Beans, dilled green	18
Chutneys:	
Tomato-apple	16
Tomato-pear	17
Cucumber pickles:	
Brined dill	9
Crosscut slices	16
Fresh-pack dill	12
Gherkins, sweet	13
Pickled fruits:	
Peaches	18
Pears	19
Watermelon	20
Relishes:	
Corn	26
Horseradish	25
Pepper-onion	25
Piccalilli	26
Sauerkraut	21

HOME FREEZING OF FRUITS AND VEGETABLES

Contents

	Page
General freezing procedures	3
What to freeze	3
Containers for freezing	4
Packing	6
Loading the freezer	7
In case of emergency	8
Refreezing	8
Points on freezing fruits	9
Before packing	9
Ways to pack	10
To keep fruit from darkening	11
Table of fruit yields	12
Directions for fruits	18
Points on freezing vegetables	29
First steps	29
Heating before packing	29
Cooling	30
Dry pack more practical	30
Table of vegetable yields	30
Directions for vegetables	36
How to use frozen fruits and vegetables	42
Fruits	42
Vegetables	43
Index	46

Home freezing of fruits and vegetables

There is no "out of season" for products of your garden and orchard—if you have a home freezer or space in a neighborhood locker plant.

Freezing is one of the simplest and least time-consuming ways to preserve foods at home. It keeps well the natural color, fresh flavor, and nutritive values of most fruits and vegetables. Frozen fruits and vegetables are ready to serve on short notice because most of the preparation they need for the table is done before freezing.

Directions are given in this bulletin for freezing many fruits and vegetables that give satisfactory products when frozen at home or in the locker plant. It is important that the directions be followed carefully, because the quality of product can vary with freshness of produce used, method of preparation and packaging, and conditions of freezing.

General freezing procedures
What to freeze

Freezing is not necessarily recommended as the preferred way for preserving all products listed in this bulletin. What to freeze must be decided on the basis of family needs and desires, on freezer space and cost of freezer storage, and on other storage facilities available.

It may be more economical, for instance, to store some fruits and vegetables in a vegetable cellar than to freeze them. But to you freezing may be worth the extra cost because of the convenience of having the products prepared so they can be readied quickly for serving.

Costs of owning and operating a home freezer vary with the rate of turnover of foods, electricity used, costs of packaging materials, repairs, and the original price of the freezer.

Some varieties of all fruits and vegetables freeze better than others. Because growing conditions differ widely throughout the country and different varieties of fruits and vegetables are available in different localities, it is not practical to specify in this publication the varieties suitable for freezing. Write to your State extension service, experiment station, or college of agriculture for information on local varieties that give highest quality when frozen.

If you have doubts as to how well a fruit or vegetable will freeze, it would be well to test it before freezing large quantities. To test, freeze three or four packages and sample the food after freezing. This shows the effect of freezing only, not the effect of storage.

Some fruits and vegetables do not make satisfactory products when frozen. They include green onions, lettuce and other salad greens, radishes, tomatoes (except as juice or cooked). Research may provide directions later for preparing good frozen products from some of these foods.

Containers for freezing

The prime purpose of packaging is to keep food from drying out and to preserve food value, flavor, color, and pleasing texture.

All containers should be easy to seal and waterproof so they will not leak. Packaging materials must be durable and must not become so brittle at low temperatures that they crack.

To retain highest quality in frozen food, packaging materials should be moisture-vapor-proof, to prevent evaporation. Many of the packaging materials on the market for frozen food are not moisture-vapor-*proof*, but are sufficiently moisture-vapor-*resistant* to retain satisfactory quality of fruits and vegetables during storage. Glass, metal, and rigid plastic are examples of moisture-vapor-proof packaging materials. Most bags, wrapping materials, and waxed cartons made especially for freezing are moisture-vapor-resistant. Not sufficiently moisture-vapor-resistant to be suitable for packaging foods to be frozen are ordinary waxed papers, and paper cartons from cottage cheese, ice cream, and milk.

Rigid containers. Rigid containers made of aluminum, glass, plastic, tin, or heavily waxed cardboard are suitable for all packs, and especially good for liquid packs. Glass canning jars may be used for freezing most fruits and vegetables except those packed in water. Plain tin or R-enamel cans may be used for all foods, but some foods may be better packed in cans with special enamel linings: C-enamel for foods containing considerable sulfur—corn, lima beans, carrots; R-enamel for highly colored foods—beets, berries, red cherries, fruit juices, plums, pumpkin, rhubarb, squash, sweetpotatoes.

Nonrigid containers. Bags and sheets of moisture-vapor-resistant cellophane, heavy aluminum foil, pliofilm, polyethylene, or laminated papers and duplex bags consisting of various combinations of paper, metal foil, glassine, cellophane, and rubber latex are suitable for dry-packed vegetables and fruits. Bags also can be used for liquid packs.

Bags and sheets are used with or without outer cardboard cartons to protect against tearing. Bags without a protective carton are difficult to stack. The sheets may be used for wrapping such foods as corn-on-the-cob or asparagus. Some of the sheets may be heat-sealed to make a bag of the size you need. Sheets that are heat-sealing on both sides may be used as outer wraps for folding paperboard cartons.

Size. Select a size that will hold only enough of a fruit or vegetable for one meal for your family.

Shape. Rigid containers that are flat on both top and bottom stack well in a freezer. Round containers and those with flared sides or raised bottoms waste freezer space. Nonrigid containers that bulge waste freezer space.

Food can be removed easily, before it is thawed, from containers with sides that are straight from bottom to top or that flare out. Food must be partially thawed before it can be removed from containers with openings narrower than the body of the container.

Bags, sheets, and folding paperboard cartons take up little room when not in use. Rigid containers with flared sides will stack one inside the other and save space in your cupboard when not in use. Those with straight sides or narrow top openings cannot be nested.

Sealing. Care in sealing is as important as using the right container. Rigid containers usually are sealed either by pressing on or screwing on the lid. Tin cans such as are used in home canning require a sealing machine or special lids. Some rigid cardboard cartons need to have freezer tape or special wax applied after sealing to make them airtight and leakproof. Glass jars must be sealed with a lid containing composition rubber or with a lid and a rubber ring.

Most bags used for packaging can be heat-sealed or sealed by twisting and folding back the top of the bag and securing with a string, a good quality rubber or plastic band, or other sealing device available on the market. Some duplex bags are sealed by folding over a metal strip attached to the top of the bag.

Special equipment for heat-sealing bags or sheets for freezing is available on the market, or a household iron may be used. To heat-seal polyethylene or pliofilm bags or sheets used as overwraps, first place a piece of paper or heat-resistant material made especially for the purpose over the edges to be sealed. Then press with a warm iron. Regulate heat of the iron carefully—too much heat melts or crinkles the materials and prevents sealing.

As manufacturers are constantly making improvements and developing new containers it is a good idea, when you buy, to note how containers are to be sealed.

Reuse. Tin cans (with slip-top closures), glass, rigid plastic, and aluminum containers can be reused indefinitely. It is difficult to reuse aluminum foil boxes, because edges of lids and containers are folded over in sealing. Tin cans that require a sealer must be reflanged with a special attachment to a sealer before they are reused. A tin can or lid that is dented should not be used if it cannot be sealed.

Reuse of rigid cardboard cartons, unless plastic-lined, is not generally advisable because cleaning is difficult. Folding paperboard cartons used to protect an inner bag can be reused.

Cost. When you compare prices of the containers that are available in your locality, consider whether they will be reusable or not. If containers are reusable, a higher initial cost may be a saving in the long run.

Care of packaging materials. Protect packaging materials from dust and insects. Keep bags and rolls of wrapping materials that may become brittle, such as cellophane, in a place that is cool and not too dry.

Freezing accessories. Check on other items that help make packaging easier. Some containers are easier to fill if you use a stand and funnel. Special sealing irons available on the market or a regular household iron may be used for heat-sealing bags, wrappers, and some types of paper cartons. With some sealing irons, a small wooden block or box makes sealing of bags easier and quicker.

Packing

- Pack food and sirup cold into containers. Having materials cold speeds up freezing and helps retain natural color, flavor, and texture of food.

- Pack foods tightly to cut down on the amount of air in the package.

- When food is packed in bags, press air out of unfilled part of bag. Press firmly to prevent air from getting back in. Seal immediately, allowing the head space recommended for the product.

- Allow ample head space. With only a few exceptions, allowance for head space is needed between packed food and closure because food expands as it freezes. A guide to the amount of head space to allow is given in the table on the following page.

Head space to allow between packed food and closure

TYPE OF PACK	Container with wide top opening [1]		Container with narrow top opening [2]	
	Pint	Quart	Pint	Quart
Liquid pack........ (Fruit packed in juice, sugar, sirup, or water; crushed or puree; juice.)	½ inch	1 inch	¾ inch [3]	1½ inches
Dry pack [4]........ (Fruit or vegetable packed without added sugar or liquid.)	½ inch	½ inch	½ inch	½ inch

[1] This is head space for tall containers—either straight or slightly flared.
[2] Glass canning jars may be used for freezing most fruits and vegetables except those packed in water.
[3] Head space for juice should be 1½ inches.
[4] Vegetables that pack loosely, such as broccoli and asparagus, require no head space.

● Keep sealing edges free from moisture or food so that a good closure can be made. Seal carefully.

● Label packages plainly. Include name of food, date it was packed, and type of pack if food is packed in more than one form. Gummed labels, colored tape, crayons, pens, and stamps are made especially for labeling frozen food packages.

Loading the freezer

Freeze fruits and vegetables soon after they are packed. Put them in the freezer a few packages at a time as you have them ready, or keep packages in the refrigerator until all you are doing at one time are ready. Then transfer them to the home freezer or carry them in an insulated box or bag to the locker plant. Freeze at 0° F. or below.

Put no more unfrozen food into a home freezer than will freeze within 24 hours. Usually this will be about 2 or 3 pounds of food to each cubic foot of its capacity. Overloading slows down the rate of freezing, and foods that freeze too slowly may lose quality or spoil. For quickest freezing, place packages against freezing plates or coils and leave a little space between packages so air can circulate freely.

After freezing, packages may be stored close together. Store them at 0° F. or below. At higher temperatures foods lose quality much faster.

Most fruits and vegetables maintain high quality for 8 to 12 months at 0° or below; citrus fruits and citrus juices, for 4 to 6 months. Unsweetened fruits lose quality faster than those packed in sugar or sirup. Longer storage will not make foods unfit for use, but may impair quality.

It's a good idea to post a list of frozen foods near the freezer and keep it up to date. List foods as you put them in freezer, with date; check foods off list as you remove them.

In case of emergency

If power is interrupted or the freezer fails to refrigerate properly, do not open the cabinet unnecessarily. Food in a loaded cabinet usually will stay frozen for 2 days, even in summer. In a cabinet with less than half a load, food may not stay frozen more than a day.

Dry ice to prevent thawing. If the power is not to be resumed within 1 or 2 days, or if the freezer may not be back to normal operation in that time, use dry ice to keep the temperature below freezing and to prevent deterioration or spoilage of frozen food.

Twenty-five pounds of dry ice in a 10-cubic-foot cabinet should hold the temperature below freezing for 2 to 3 days in a cabinet with less than half a load and 3 to 4 days in a loaded cabinet, if dry ice is obtained quickly following interruption of power. Move any food stored in a freezing compartment of a freezer to the storage compartment. Place dry ice on boards or heavy cardboard on top of packages. Open freezer only when necessary. Don't handle dry ice with bare hands; it can cause burns. When using dry ice, room should be ventilated. If you can't get dry ice, try to locate a locker plant and move food there in insulated boxes.

Refreezing

Occasionally, frozen foods are partially or completely thawed before it is discovered that the freezer is not operating.

The basis for safety in refreezing foods is the temperature at which thawed foods have been held and the length of time they were held after thawing.

You may safely refreeze frozen foods that have thawed if they still contain ice crystals or if they are still cold—about 40° F.—and have been held no longer than 1 or 2 days at refrigerator temperature after thawing. In general, if a food is safe to eat, it is safe to refreeze.

Even partial thawing and refreezing reduce quality of fruits and vegetables. Foods that have been frozen and thawed require the same care as foods that have never been frozen. Use refrozen foods as soon as possible to save as much of their eating quality as you can.

Points on freezing fruits

Most fruits can be frozen satisfactorily, but the quality of the frozen product will vary with the kind of fruit, stage of maturity, and type of pack. Pointers on selecting fruit properly are given in the directions and must be followed carefully to be sure of a good frozen product.

Generally, flavor is well retained by freezing preservation. Texture may be somewhat softer than that of fresh fruit. Some fruits require special treatment when packed to make them more pleasing in color, texture, or flavor after thawing. Most fruits are best frozen soon after harvesting. Some, such as peaches and pears, may need to be held a short time to ripen.

Before packing

All fruits need to be washed in cold water. Wash a small quantity at a time to save undue handling, which may bruise delicate fruits such as berries. A perforated or wire basket is useful. Lift washed fruits out of the water and drain thoroughly. Don't let the fruit stand in the water—some lose food value and flavor that way and some get water-soaked.

In general, fruit is prepared for freezing in about the same way as for serving. Large fruits generally make a better product if cut in pieces or crushed before freezing. Many fruits can be frozen successfully in several forms. Good parts of less perfect fruit are suitable for crushed or pureed packs.

Peel, trim, pit, and slice fruit following the directions on pages 18 to 28. It is best to prepare enough fruit for only a few containers at one time, especially those fruits that darken rapidly. Two or three quarts is a good quantity to work with.

If directions call for fruit to be crushed, suit the method of crushing to the fruit. For soft fruits, a wire potato masher, pastry fork, or slotted spoon may be used; if fruits are firm they may be crushed more easily with a food chopper. For making purees a colander, food press, or strainer is useful.

Use equipment of aluminum, earthenware, enameled ware, glass, nickel, stainless steel, or good-quality tinware. Do not use galvanized ware in direct contact with fruit or fruit juices because the acid in fruit dissolves zinc, which is poisonous.

Metallic off-flavors may result from the use of iron utensils, chipped enameled ware, or tinware that is not well tinned.

Ways to pack

Most fruits have better texture and flavor if packed in sugar or sirup. Some may be packed without sweetening.

In the directions for freezing, three ways of packing are given for fruits whole or in pieces—sirup pack, sugar pack, and unsweetened pack. Directions are also given for packing crushed fruits, purees, and fruit juices.

Your selection of the way to pack the fruit will depend on the intended use. Fruits packed in a sirup are generally best for dessert use; those packed in dry sugar or unsweetened are best for most cooking purposes because there is less liquid in the product.

Even though unsweetened packs generally yield a lower quality product than packs with sugar, directions in this publication include unsweetened packs whenever they are satisfactory, because they are often needed for special diets. Some fruits, such as gooseberries, currants, cranberries, rhubarb, and figs, give as good quality packs without as with sugar.

Sirup pack. A 40-percent sirup is recommended for most fruits. For some mild-flavored fruits lighter sirups are desirable to prevent masking of flavor. Heavier sirups may be needed for very sour fruits.

In the directions for each fruit, sirups are called for according to the percentage of sugar in the sirup. Below is a master recipe from which any of the sirups can be made. It takes one-half to two-thirds cup of sirup for each pint package of fruit.

Sirups for use in freezing fruits

Type of sirup	Sugar [1] Cups	Water Cups	Yield of sirup Cups
30-percent sirup	2	4	5
35-percent sirup	2½	4	5⅓
40-percent sirup	3	4	5½
50-percent sirup	4¾	4	6½
60-percent sirup	7	4	7¾
65-percent sirup	8¾	4	8⅔

[1] In general, up to one-fourth of the sugar may be replaced by corn sirup. A larger proportion of corn sirup may be used if a very bland, light-colored type is selected.

Dissolve sugar in cold or hot water. If hot water is used, cool sirup before using. Sirup may be made up the day before and kept cold in the refrigerator.

When packing fruit into containers be sure the sirup covers the fruit, so that the top pieces will not change in color and flavor. To keep the fruit under the sirup, place a small piece of crumpled parchment paper or other water-resistant wrapping material on top and press fruit down into sirup before closing and sealing the container.

Sugar pack. Cut fruit into a bowl or shallow pan. Sprinkle the sugar (quantity needed given in the directions for each fruit) over the fruit. To mix, use a large spoon or pancake turner. Mix gently until juice is drawn out and sugar is dissolved.

Put fruit and juice into containers. Place a small piece of crumpled parchment paper or other water-resistant wrapping material on top to hold fruit down in juice. Close and seal the container.

Unsweetened pack. Pack prepared fruit into containers, without added liquid or sweetening, or cover with water containing ascorbic acid. Or pack crushed or sliced fruit in its own juice without sweetening. Press fruit down into juice or water with a small piece of crumpled parchment paper as for sirup and sugar pack. Close and seal containers.

To keep fruit from darkening

Some fruits darken during freezing if not treated to retard darkening. Directions for such fruits list antidarkening treatment as part of the freezing preparation. Several types of antidarkening treatments are used because all fruits are not protected equally well by all treatments.

Ascorbic acid. For most of the fruits that need antidarkening treatment, ascorbic acid (vitamin C) may be used. This is very effective in preserving color and flavor of fruit and adds nutritive value.

Ascorbic acid in crystalline form is available at drug stores and at some locker plants, in various sized containers from 25 to 1,000 grams. (Crystalline ascorbic acid may be obtained also in powdered form.) One teaspoon weighs about 3 grams; thus there are approximately 8 teaspoons of ascorbic acid in a 25-gram container. In the recipes, amounts of crystalline ascorbic acid are given in teaspoons.

Ascorbic acid tablets can be used but are more expensive and more difficult to dissolve than the crystalline form. Also filler in the tablets may make the sirup cloudy. The amount of ascorbic acid in tablets is usually expressed in milligrams. Below are amounts needed in milligrams if tablets are used in place of crystalline ascorbic acid:

Crystalline	Tablets
⅛ teaspoon	375 milligrams
¼ teaspoon	750 milligrams
½ teaspoon	1,500 milligrams
¾ teaspoon	2,250 milligrams
1 teaspoon	3,000 milligrams

To use, dissolve ascorbic acid in a little cold water. If using tablets, crush them so they will dissolve more easily.

- *In sirup pack.* Add the dissolved ascorbic acid to the cold sirup shortly before using. Stir it in gently so you won't stir in air. Solutions of ascorbic acid should be made up as needed. Keep sirup in refrigerator until used.

- *In sugar pack.* Sprinkle the dissolved ascorbic acid over the fruit just before adding sugar.

- *In unsweetened pack.* Sprinkle the dissolved ascorbic acid over the fruit and mix thoroughly just before packing. If fruit is packed in water, dissolve the ascorbic acid in the water.

- *In fruit juices.* Add ascorbic acid directly to the juice. Stir only enough to dissolve ascorbic acid.

- *In crushed fruits and fruit purees.* Add dissolved ascorbic acid to the fruit preparation and mix.

Ascorbic acid mixtures. There are on the market special antidarkening preparations—usually made of ascorbic acid mixed with sugar or with sugar and citric acid. If you use one of these, follow the manufacturer's directions. In these mixtures ascorbic acid is usually the important active ingredient. Because of its dilution with other materials, ascorbic acid purchased in these forms may be more expensive than the pure ascorbic acid.

Citric acid, lemon juice. For a few fruits citric acid or lemon juice (which contains both citric acid and ascorbic acid) makes a suitable antidarkening agent. However, neither is as effective as ascorbic acid. Citric acid or lemon juice in the large quantities needed in some cases would mask the natural fruit flavors or make the fruits too sour.

Citric acid in crystalline or powdered form is available at drugstores and some locker plants. When using citric acid, dissolve it in a little cold water before adding to the fruit according to directions for that fruit.

Steam. For some fruits steaming for a few minutes before packing is enough to control darkening.

Table of fruit yields

The following table will help you figure how much frozen fruit you can get from a given quantity of fresh fruit and will help in making cost comparisons.

The number of pints of frozen food you can get depends upon the quality, variety, maturity, and size of the fruit—and whether it is frozen whole or in halves, in slices, in cubes, or in balls.

Approximate yield of frozen fruits from fresh

FRUIT	FRESH, AS PURCHASED OR PICKED	FROZEN
Apples	1 bu. (48 lb.)	32 to 40 pt.
	1 box (44 lb.)	29 to 35 pt.
	1 ¼ to 1½ lb.	1 pt.
Apricots	1 bu. (48 lb.)	60 to 72 pt.
	1 crate (22 lb.)	28 to 33 pt.
	⅔ to 4/5 lb.	1 pt.
Berries [1]	1 crate (24 qt.)	32 to 36 pt.
	1⅓ to 1½ pt.	1 pt.
Cantaloups	1 dozen (28 lb.)	22 pt.
	1 to 1¼ lb.	1 pt.
Cherries, sweet or sour	1 bu. (56 lb.)	36 to 44 pt.
	1¼ to 1½ lb.	1 pt.
Cranberries	1 box (25 lb.)	50 pt.
	1 peck (8 lb.)	16 pt.
	½ lb.	1 pt.
Currants	2 qt. (3 lb.)	4 pt.
	¾ lb.	1 pt.
Peaches	1 bu. (48 lb.)	32 to 48 pt.
	1 lug box (20 lb.)	13 to 20 pt.
	1 to 1½ lb.	1 pt.
Pears	1 bu. (50 lb.)	40 to 50 pt.
	1 western box (46 lb.)	37 to 46 pt.
	1 to 1¼ lb.	1 pt.
Pineapple	5 lb.	4 pt.
Plums and prunes	1 bu. (56 lb.)	38 to 56 pt.
	1 crate (20 lb.)	13 to 20 pt.
	1 to 1½ lb.	1 pt.
Raspberries	1 crate (24 pt.)	24 pt.
	1 pt.	1 pt.
Rhubarb	15 lb.	15 to 22 pt.
	⅔ to 1 lb.	1 pt.
Strawberries	1 crate (24 qt.)	38 pt.
	⅔ qt.	1 pt.

[1] Includes blackberries, blueberries, boysenberries, dewberries, elderberries, gooseberries, huckleberries, loganberries, and youngberries.

Strawberries . . . packed in sugar

Pride of the freezer are strawberries—sliced, sweetened with dry sugar, and frozen. For other fruits packed in sugar, follow the general steps shown here. A pint plastic box is the container illustrated, but other types of containers (p. 4) may also be used.

- Select firm, ripe strawberries—about ⅔ quart fresh berries are needed for each pint frozen.

9116D

- Wash berries a few at a time in cold water. Lift berries gently out of water and drain.

9117D

- Remove hulls; then slice berries into a bowl or shallow pan.

9119D

● Sprinkle sugar over berries—
¾ cup to each quart (1⅓
pounds) berries. Turn berries
over and over until sugar is
dissolved and juice is formed.

● Pack berries in container,
leaving ½-inch head space in
the wide-mouth pint box. Place
a small piece of crumpled
parchment paper on top of
berries. Press berries down into juice.

● Press lid on firmly to seal.
Be sure the seal is watertight.

● Label package with name of
fruit and date frozen. Freeze;
then store at 0° F. or below.

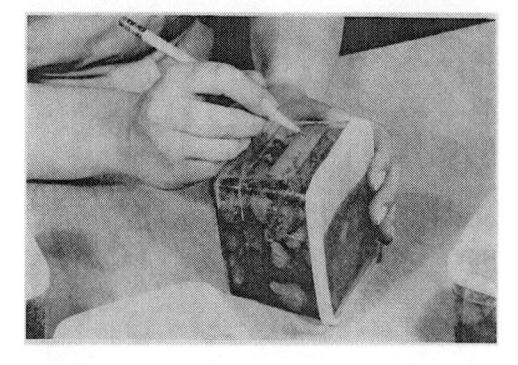

Home Freezing of Fruits and Vegetables (15) 135

Peaches . . . packed in sirup

Peaches packed in either sirup or sugar make an excellent frozen product. Sliced peaches are shown being packed in sirup. A pint glass freezer jar is used here, but other sizes and types of containers (p. 4) are suitable.

Follow these general directions for packing other fruits in sirup. Vary the sirup as called for in the directions for each fruit.

Make up sirup ahead of time so it will be ready and cold when you need it. Peaches are best packed in a 40-percent sirup—3 cups of sugar to 4 cups of water. This amount makes about 5½ cups of sirup. You need about ⅔ cup of sirup for each pint container of peaches. For details of sirup making, see page 10.

For frozen peaches with better color and flavor, add ascorbic acid to the cold sirup as described on pages 11 and 12. For peaches, use ½ teaspoon crystalline ascorbic acid to each quart of sirup.

● Select mature peaches that are firm-ripe, with no green color in the skins. Allow 1 to 1½ pounds fresh peaches for each pint to be frozen. Wash them carefully and drain.

286A

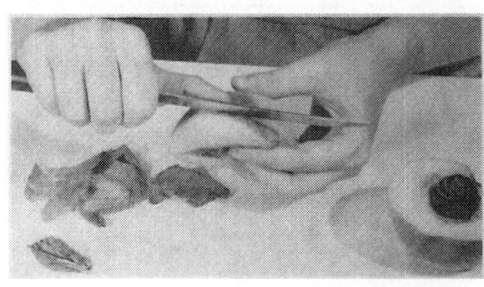

● Pit peaches, and peel them by hand for the best-looking product. Peaches peel more quickly if they are dipped first in boiling water, then cold—but have ragged edges after thawing.

287A

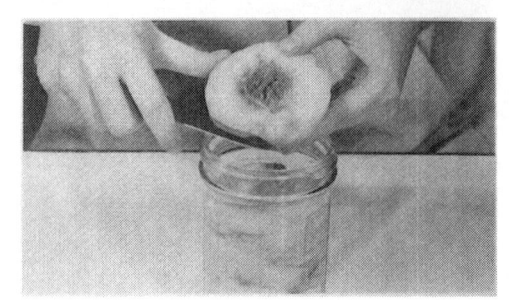

● Pour about ½ cup cold sirup into each pint container. Slice peaches directly into container.

288A

● Add sirup to cover peaches. Leave ½-inch head space at top of wide-mouth pint containers such as these, to allow for the expansion of the fruit during freezing.

289A

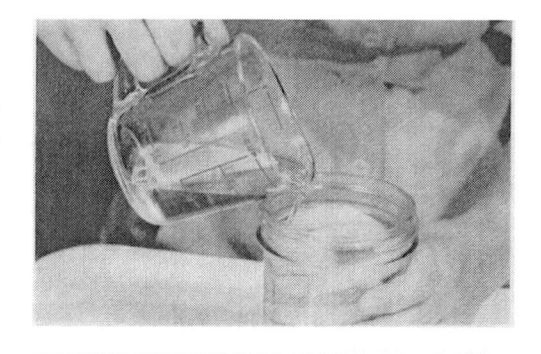

● Put a small piece of crumpled parchment paper on top of fruit to keep peaches down in the sirup. Sirup should always cover fruit to keep top pieces from changing color and flavor.

291A

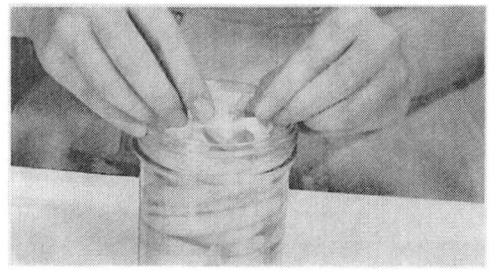

● Wipe all sealing edges clean for a good seal. Screw lid on tight. Label with name of fruit and date of freezing.

292A

● Put sealed containers in the coldest part of freezer or locker. Leave a little space between containers so air can circulate freely. After fruit is frozen, store at 0° F. or below.

293A

Home Freezing of Fruits and Vegetables (17) 137

Directions for fruits

Apples, slices

Sirup pack is preferred for apples to be used for fruit cocktail or uncooked dessert. Apples packed in sugar or frozen unsweetened are good for pie making. For better quality, apple slices need to be treated to prevent darkening.

Select full-flavored apples that are crisp and firm, not mealy in texture.[1] Wash, peel, and core. Slice medium apples into twelfths, large ones into sixteenths.

Pack in one of the following ways:

Sirup pack. Use 40-percent sirup (p. 10). For a better quality frozen product add ½ teaspoon crystalline ascorbic acid to each quart of sirup.

Slice apples directly into cold sirup in container, starting with ½ cup sirup to a pint container. Press fruit down in containers and add enough sirup to cover.

Leave head space (p. 7). Seal and freeze.

Sugar pack. To prevent darkening of apples during preparation, slice them into a solution of 2 tablespoons salt to a gallon of water. Hold in this solution no more than 15 to 20 minutes. Drain.

To retard darkening, place slices in a single layer in steamer; steam 1½ to 2 minutes, depending on thickness of slice. Cool in cold water; drain.

Over each quart (1¼ pounds) of apple slices sprinkle evenly ½ cup sugar and stir.

Pack apples into containers and press fruit down, leaving head space (p. 7). Seal and freeze.

Unsweetened pack. Follow directions for sugar pack, omitting sugar.

Applesauce

Select full-flavored apples. Wash apples, peel if desired, core, and slice. To each quart of apple slices add ⅓ cup water; cook until tender. Cool and strain if necessary. Sweeten to taste with ¼ to ¾ cup sugar for each quart (2 pounds) of sauce.

Pack into containers, leaving head space (p. 7). Seal and freeze.

Apricots

● **Halves and slices.** The sirup pack is preferred for fruit to be served uncooked; the sugar pack for apricots to be used for pies or other cooked dishes.

Treatment to prevent darkening is necessary for a satisfactory product if apricots are packed in sugar. Such treatment also improves the quality of apricots packed in sirup.

Select firm, ripe, uniformly yellow apricots. Sort, wash, halve, and pit. Peel and slice if desired.

---

[1] To firm soft apples that are to be used in cooking or baking after freezing:
Hold sliced apples for 5 to 20 minutes in a solution made from 1 teaspoon calcium chloride (U.S.P. grade) or 2 tablespoons calcium lactate (U.S.P. grade) to each quart of water. The softer the apples the longer the time they should be held in the solution.
Apples differ with variety, stage of ripeness, and the region in which they are grown. Make a trial with a few packages. After freezing, boil apple slices a few minutes to test firmness.

If apricots are not peeled, heat them in boiling water ½ minute to keep skins from toughening during freezing. Then cool in cold water and drain.

Pack into containers in one of the following ways:

Sirup pack. Use 40-percent sirup (p. 10). For a better quality frozen product, add ¾ teaspoon crystalline ascorbic acid to each quart of sirup.

Pack apricots directly into containers. Cover with sirup, leaving head space (p. 7). Seal and freeze.

Sugar pack. Before combining apricots with sugar give the fruit the following treatment to prevent darkening:

Dissolve ¼ teaspoon crystalline ascorbic acid in ¼ cup cold water and sprinkle over 1 quart (⅞ pound) of fruit.

Mix ½ cup sugar with each quart of fruit. Stir until sugar is dissolved. Pack apricots into containers and press down until fruit is covered with juice, leaving head space (p. 7). Seal and freeze.

● **Crushed or puree.** Select fully ripe fruit. For crushed apricots, dip in boiling water for ½ minute and cool in cold water. Peel the apricots. Pit and crush them coarsely.

For puree, pit and quarter the apricots. Press through a sieve; or heat to boiling point in just enough water to prevent scorching and then press through a sieve.

With each quart (2 pounds) of prepared apricots mix 1 cup sugar. For a better product, add ¼ teaspoon crystalline ascorbic acid dissolved in ¼ cup of water to the fruit just before adding the sugar.

Pack into containers, leaving head space (p. 7). Seal and freeze.

Avocados

● **Puree.** Avocados are best frozen as puree—unsweetened for salads and sandwiches, sweetened for ice cream and milk shakes. Avocados are not satisfactorily frozen whole or sliced.

Select avocados that are soft ripe—not hard or mushy—with rinds free from dark blemishes. Peel the fruit, cut in half, and remove the pit. Mash the pulp.

Pack in one of the following ways:

Unsweetened pack. For a better quality product add ⅛ teaspoon crystalline ascorbic acid to each quart of puree.

Pack into containers, leaving head space (p. 7). Seal and freeze.

Sugar pack. Mix 1 cup sugar with 1 quart (2 pounds) of puree. Pack into containers, leaving head space (p. 7). Seal and freeze.

Blackberries, boysenberries, dewberries, loganberries, youngberries

● **Whole.** The sirup pack is preferred for berries to be served uncooked. The sugar pack or the unsweetened pack is satisfactory for berries to be used for cooked products such as pie or jam.

(Continued on page 20)

Blackberries—continued

Select firm, plump, fully ripe berries with glossy skins. Green berries may cause off-flavor.

Sort and remove any leaves and stems. Wash and drain.

Use one of the three following packs:

Sirup pack. Pack berries into containers and cover with cold 40- or 50-percent sirup (p. 10), depending on the sweetness of the fruit. Leave head space (p. 7). Seal and freeze.

Sugar pack. To 1 quart (1⅓ pounds) berries, add ¾ cup sugar. Turn berries over and over until most of the sugar is dissolved. Fill containers, leaving head space (p. 7). Seal and freeze.

Unsweetened pack. Pack berries into containers, leaving head space (p. 7). Seal and freeze.

● **Crushed or puree.** Prepare for packing in same way as for whole berries. Then crush. Or press through a sieve for puree.

To each quart (2 pounds) of crushed berries or puree add 1 cup sugar. Stir until sugar is dissolved. Pack into containers, leaving head space (p. 7). Seal and freeze.

Blueberries, elderberries, huckleberries

● **Whole.** The sirup pack is preferred for berries to be served uncooked. Berries frozen unsweetened are satisfactory for cooking.

Select full-flavored, ripe berries all about the same size, preferably with tender skins. Sort, wash, and drain. If desired, steam for 1 minute and cool immediately. Preheating in steam tenderizes skin and makes a better flavored product.

Use one of the following packs:

Sirup pack. Pack berries into containers and cover with cold 40-percent sirup (p. 10). Leave head space (p. 7). Seal and freeze.

Unsweetened pack. Pack berries into containers, leaving head space (p. 7). Seal and freeze.

● **Crushed or puree.** Select fully ripened berries. Sort, wash, and drain. Crush, or press berries through a fine sieve for puree.

To 1 quart (2 pounds) crushed berries or puree, add 1 to 1⅛ cups sugar, depending on tartness of fruit. Stir until sugar is dissolved. Pack into containers, leaving head space (p. 7). Seal and freeze.

Cherries, sour

● **Whole.** Sirup pack is best for cherries to be served uncooked. Sugar pack is preferable for cherries to be used for pies or other cooked products. (Directions for packing sweet with sour cherries are on p. 21.)

Select bright-red, tree-ripened cherries. Stem, sort, and wash thoroughly. Drain and pit.

Use one of the following packs:

Sirup pack. Pack cherries into containers and cover with cold 60- or 65-percent sirup (p. 10), depending on tartness of the cherries. Leave head space (p. 7). Seal and freeze.

Sugar pack. To 1 quart (1⅓ pounds) cherries add ¾ cup sugar. Mix until sugar is dissolved. Pack into containers, leaving head space (p. 7). Seal and freeze.

● **Crushed.** Prepare for packing as for whole sour cherries. Crush coarsely.

To 1 quart (2 pounds) fruit add 1 to 1½ cups sugar, depending on sweetness desired. Mix thoroughly until sugar is dissolved. Pack into containers, leaving head space (p. 7). Seal and freeze.

● **Puree.** Select and prepare for packing same as for whole cherries. Then crush cherries, heat to boiling point, cool, and press through a sieve.

To 1 quart (2 pounds) fruit puree add ¾ cup sugar. Pack puree into containers, leaving head space (p. 7). Seal and freeze.

● **Juice.** Select and prepare as for whole sour cherries. Then crush cherries, heat slightly to start flow of juice, and strain juice through a jelly bag. Cool, let stand overnight, and pour off clear juice for freezing. Or juice may be packed as soon as it cools, then strained when it is thawed for serving.

Sweeten with 1½ to 2 cups sugar to each quart of juice or pack without added sugar. Pour into containers, leaving head space (p. 7). Seal and freeze.

Cherries, sweet

● **Whole.** Sweet cherries should be prepared quickly to avoid color and flavor changes. Red varieties are best for freezing.

Select well-colored, tree-ripened fruit with a sweet flavor. Sort, stem, wash, and drain. Remove pits if desired; they tend to give an almond-like flavor to the fruit.

Pack cherries into containers. Cover with cold 40-percent sirup (p. 10) to which has been added ½ teaspoon crystalline ascorbic acid to the quart. Leave head space (p 7). Seal and freeze.

With sour cherries. Use half sweet cherries, half sour. Pack as above using 50-percent sirup (p. 10). Ascorbic acid may be added, but is not essential as it is for sweet cherries alone.

● **Crushed.** Prepare cherries as for freezing whole. Remove pits and crush cherries coarsely.

To each quart (2 pounds) of crushed fruit add 1½ cups sugar and ¼ teaspoon crystalline ascorbic acid. Mix well. Pack into containers, leaving head space (p. 7). Seal and freeze.

● **Juice.** Frozen sweet cherry juice may lack flavor and tartness. For a tastier product, add some sour cherry juice—either before freezing or after thawing.

Select well-colored, tree-ripened fruit. Sort, stem, wash, and drain. Remove pits and crush.

For red cherries, heat slightly (to 165° F.) to start flow of juice. Do not boil. Extract juice in a jelly bag.

For white cherries, extract juice without heating. Then warm juice (to 165° F.) in a double boiler or over low heat.

(Continued on page 22)

Cherries, sweet—continued

Juice—continued. For either red or white cherry juice, cool the juice, let stand overnight, and pour off clear juice for freezing. Or pack the juice as soon as it cools; then strain after thawing for serving.

Sweeten with 1 cup sugar to each quart of juice, or pack without adding sugar. Pour into container, leaving head space (p. 7). Seal and freeze.

Coconut, fresh

Shred coconut meat or put it through a food chopper. Pack into containers and cover with the coconut milk. Leave head space (p. 7). Seal and freeze.

Cranberries

● **Whole.** Choose firm, deep-red berries with glossy skins. Stem and sort. Wash and drain.

Unsweetened pack. Pack into containers without sugar. Leave head space (p. 7). Seal and freeze.

Sirup pack. Pack into containers. Cover with cold 50-percent sirup (p. 10). Leave head space (p. 7). Seal and freeze.

● **Puree.** Prepare cranberries as for freezing whole. Add 2 cups water to each quart (1 pound) of berries. Cook until skins have popped. Press through a sieve.

Add sugar to taste, about 2 cups for each quart (2 pounds) of puree. Pack into containers, leaving head space (p. 7). Seal and freeze.

Currants

● **Whole.** Select plump, fully ripe bright-red currants. Wash in cold water and remove stems.

Pack in any of the following ways:

Unsweetened pack. Pack into containers, leaving head space (p. 7). Seal and freeze.

Sirup pack. Pack into containers and cover currants with cold 50-percent sirup (p. 10), leaving head space (p. 7). Seal and freeze.

Sugar pack. To each quart (1⅛ pounds) of fruit add ¾ cup sugar. Stir until most of the sugar is dissolved. Pack currants into containers, leaving head space (p. 7). Seal and freeze.

● **Crushed.** Prepare as directed for whole currants. Crush.

To 1 quart (2 pounds) crushed currants add 1⅛ cups sugar. Mix until sugar is dissolved. Pack into containers, leaving head space (p. 7). Seal and freeze.

● **Juice.** For use in beverages, select as directed for whole currants. For use in jelly making, mix slightly underripe and ripe fruit. Wash in cold water and remove stems. Crush currants and warm (to 165° F.) over low heat to start flow of juice. Do not boil. Press hot fruit in jelly bag to extract juice. Cool.

Sweeten with ¾ to 1 cup sugar to each quart of juice, or pack without adding sugar. Pour into containers, leaving head space (p. 7). Seal and freeze.

Dates

Select dates with good flavor and tender texture. Wash and slit to remove pits. Leave whole, or press through a sieve for puree.

Pack into containers, leaving head space (p. 7). Seal and freeze.

Figs

● **Whole or sliced.** Select tree-ripened soft-ripe fruit. Make sure figs have not become sour in the center. Sort, wash, and cut off stems. Peel if desired. Slice or leave whole.

Use one of the following packs:

Sirup pack. Use 35-percent sirup (p. 10). For a better product add ¾ teaspoon crystalline ascorbic acid or ½ cup lemon juice to each quart of sirup. Pack figs into containers and cover with cold sirup, leaving head space (p. 7). Seal and freeze.

Unsweetened pack. Pack into containers, leaving head space (p. 7). Cover with water or not as desired. If water is used, crystalline ascorbic acid may be added to retard darkening of light-colored figs—¾ teaspoon to each quart of water. Leave head space (p. 7). Seal and freeze.

● **Crushed.** Prepare figs as directed for freezing whole or sliced. Crush them coarsely.

With 1 quart (1½ pounds) fruit, mix ⅔ cup sugar. For a product of better quality add ¼ teaspoon crystalline ascorbic acid to each quart of fruit. Pack figs into containers, leaving head space (p. 7). Seal and freeze.

Fruit cocktail

Use any combination of fruits desired . . . sliced or cubed peaches or apricots, melon balls, orange or grapefruit sections, whole seedless grapes, Bing cherries, or pineapple wedges.

Pack into containers; cover with cold 30- or 40-percent sirup (p. 10), depending on fruits used. Leave head space (p. 7). Seal and freeze.

Gooseberries

Whole gooseberries may be frozen with sirup or without sweetening. For use in pie or preserves, the unsweetened pack is better.

Choose fully ripe berries if freezing for pie—berries a little underripe for jelly making. Sort, remove stems and blossom ends, and wash.

Unsweetened pack. Pack into containers without sugar. Leave head space (p. 7). Seal and freeze.

Sirup pack. Pack into containers. Cover with 50-percent sirup (p. 10). Leave head space (p. 7). Seal and freeze.

Grapefruit, oranges

● **Sections or slices.** Select firm tree-ripened fruit heavy for its size and free from soft spots. Wash and peel. Divide fruit into sections, removing all membranes and seeds. Slice oranges if desired. For grapefruit with many seeds, cut fruit in half, remove seeds; cut or scoop out sections.

(Continued on page 24)

Grapefruit—continued

Sections or slices—continued. Pack fruit into containers. Cover with cold 40-percent sirup (p. 10) made with excess fruit juice and water if needed. For better quality, add ½ teaspoon crystalline ascorbic acid to a quart of sirup. Leave head space (p. 7). Seal and freeze.

● **Juice.** Select fruit as directed for sections. Squeeze juice from fruit, using squeezer that does not press oil from rind.

Sweeten with 2 tablespoons sugar for each quart of juice, or pack without sugar. For better quality, add ¾ teaspoon crystalline ascorbic acid for each gallon of juice. Pour juice into containers immediately. To avoid development of off-flavors, pack juice in glass jars or citrus-enamel tin cans, if available. Leave head space (p. 7). Seal and freeze.

Grapes

● **Whole or halves.** Grapes are best frozen with sirup, but grapes to be used for juice or jelly can be frozen without sweetening.

Select firm-ripe grapes with tender skins and full color and flavor. Wash and stem. Leave seedless grapes whole; cut table grapes with seeds in half and remove seeds.

Unsweetened pack. Pack into containers without sweetening. Leave head space (p. 7). Seal and freeze.

Sirup pack. Pack into containers and cover with cold 40-percent sirup (p. 10). Leave head space (p. 7). Seal and freeze.

● **Puree.** Grapes may be frozen as puree with sugar added. The puree may develop a gritty texture because of tartrate crystals. The crystals disappear when puree is heated.

Wash, stem, and crush the grapes. Heat to boiling. Drain off free juice and freeze or can it separately. Cool the crushed grapes and press them through a sieve.

To 1 quart (2 pounds) puree add ½ cup sugar. Pack into containers, leaving head space (p. 7). Seal and freeze.

● **Juice.** For beverages, select as for whole grapes. For jelly making, select as recommended in specific jelly recipe.

Wash, stem, and crush grapes. Strain them through a jelly bag. Let juice stand overnight in refrigerator or other cool place while sediment sinks to bottom. Pour off clear juice for freezing.

Pour juice into containers, leaving head space (p. 7). Seal and freeze.

If tartrate crystals form in frozen juice, they may be removed by straining the juice after it thaws.

Melons—cantaloup, crenshaw, honeydew, Persian, watermelon

● **Slices, cubes, or balls.** Select firm-fleshed, well-colored, ripe melons. Cut in half, remove seeds, and peel. Cut melons into slices, cubes, or balls. Pack into containers and cover with cold 30-percent sirup (p. 10). Leave head space (p. 7). Seal and freeze.

● **Crushed.** Prepare melons, except watermelon, as for freezing in slices, cubes, or balls. Then crush them. If a food chopper is used for crushing, use the coarse knife.

Add 1 tablespoon sugar to each quart of crushed fruit if desired. Stir until sugar is dissolved. Pack melon into containers, leaving head space (p. 7). Seal and freeze.

Nectarines

● **Halves, quarters, or slices.** Choose fully ripe, well-colored, firm nectarines. Overripe fruit may take on a disagreeable flavor in frozen storage.

Sort, wash, and pit the fruit. Peel if desired. Cut in halves, quarters, or slices.

Cut fruit directly into cold 40-percent sirup (p. 10), starting with ½ cup for each pint container. For a better product add ½ teaspoon crystalline ascorbic acid to each quart of sirup. Press fruit down and add sirup to cover, leaving head space (p. 7). Seal and freeze.

● **Puree.** Prepare same as peach puree (see below).

Peaches

● **Halves and slices.** Peaches in halves and slices have better quality when packed in sirup or with sugar, but a water pack will serve if sweetening is not desired.

Select firm, ripe peaches with no green color in the skins.

Sort, wash, pit, and peel. For a better product, peel peaches without a boiling-water dip. Slice if desired.

Sirup pack. Use 40-percent sirup (p. 10). For a better quality product, add ½ teaspoon crystalline ascorbic acid for each quart of sirup.

Put peaches directly into cold sirup in container—starting with ½ cup sirup to a pint container. Press fruit down and add sirup to cover, leaving head space (p. 7). Seal and freeze.

Sugar pack. To each quart (1⅓ pounds) of prepared fruit add ⅔ cup sugar and mix well. To retard darkening, sprinkle ascorbic acid dissolved in water over the peaches before adding sugar. Use ¼ teaspoon crystalline ascorbic acid in ¼ cup cold water to each quart of fruit.

Pack into containers, leaving head space (p. 7). Seal and freeze.

Water pack. Pack peaches into containers and cover with cold water containing 1 teaspoon crystalline ascorbic acid to each quart of water. Leave head space (p. 7). Seal and freeze.

● **Crushed or puree.** To loosen skins, dip peaches in boiling water ½ to 1 minute. The riper the fruit the less scalding needed. Cool in cold water, remove skins, and pit.

Crush peaches coarsely. Or, for puree, press through a sieve, or heat pitted peaches 4 minutes in just enough water to prevent scorching and then press through a sieve.

With each quart (2 pounds) of crushed or pureed peaches mix 1 cup sugar. For better quality, add ⅛ teaspoon crystalline ascorbic acid to each quart of fruit.

Pack into containers, leaving head space (p. 7). Seal and freeze.

Pears

● **Halves or quarters.** Select pears that are well ripened and firm but not hard. Wash fruit in cold water. Peel, cut in halves or quarters, and remove cores.

Heat pears in boiling 40-percent sirup (p. 10) for 1 to 2 minutes, depending on size of pieces. Drain and cool.

Pack pears into containers and cover with cold 40-percent sirup (p. 10). For a better product, add ¾ teaspoon crystalline ascorbic acid to a quart of cold sirup. Leave head space (p. 7). Seal and freeze.

● **Puree.** Select well-ripened pears, firm but not hard or gritty. Peel or not as desired, but do not dip in boiling water to remove skins. Prepare and pack as for peach puree (see directions on p. 25).

Persimmons

● **Puree (cultivated and native varieties).** Select orange-colored, soft-ripe persimmons. Sort, wash, peel, and cut into sections. Press the fruit through a sieve.

To each quart of persimmon puree add ⅛ teaspoon crystalline ascorbic acid or 1½ teaspoons crystalline citric acid to help prevent darkening and flavor loss.

Persimmon puree made from native varieties needs no sugar. Puree made from cultivated varieties may be packed with or without sugar.

Unsweetened pack. Pack unsweetened puree into containers. Leave head space (p. 7). Seal and freeze.

Sugar pack. Mix 1 cup sugar with 1 quart (2 pounds) puree and pack into containers. Leave head space (p. 7). Seal and freeze.

Pineapple

Select firm, ripe pineapple with full flavor and aroma. Pare and remove core and eyes. Slice, dice, crush, or cut the pineapple into wedges or sticks.

Unsweetened pack. Pack fruit tightly into containers without sugar. Leave head space (p. 7). Seal and freeze.

Sirup pack. Pack fruit tightly into containers. Cover with 30-percent sirup made with pineapple juice, if available, or with water (p. 10). Leave head space (p. 7). Seal and freeze.

Plums and prunes

● **Whole, halves, or quarters.** Frozen plums and prunes are very good for use in pies and jams, or in salads and desserts. The unsweetened pack is preferred for plums to be used for jams.

Choose firm tree-ripened fruit of deep color. Sort and wash. Leave whole or cut in halves or quarters. Pack in one of the following ways:

Unsweetened pack. Pack whole fruit into containers, leaving head space (p. 7). Seal and freeze.

To serve uncooked, dip frozen fruit in cold water for 5 to 10 seconds, remove skins, and cover with 40-percent sirup to thaw.

Sirup pack. Pack cut fruit into containers. Cover fruit with cold 40- or 50-percent sirup, depending on tartness of fruit (p. 10). For improved quality, add ½ teaspoon crystalline ascorbic acid to a quart of sirup. Leave head space (p. 7). Seal and freeze.

● **Puree.** Select fully ripe fruit. Wash, cut in halves, and remove pits. Puree may be prepared from unheated or heated fruit, depending on softness of fruit.

To prepare puree from unheated fruit, press raw fruit through a sieve. For better quality, add either ¼ teaspoon crystalline ascorbic acid or ½ tablespoon crystalline citric acid to each quart (2 pounds) of puree.

To prepare puree from heated fruit, add 1 cup water for each 4 quarts (4 pounds) of fruit. Bring to a boil, cook 2 minutes, cool, and press through a sieve.

With each quart (2 pounds) of puree, mix ½ to 1 cup sugar, depending on tartness of fruit. Pack into containers, leaving head space (p. 7). Seal and freeze.

● **Juice.** For juice to be served in beverages, select plums as for puree. For juice to be used for jelly making, select as recommended in specific jelly recipe. Wash plums, then simmer until soft in enough water to barely cover. Strain through a jelly bag. Cool.

If desired, sweeten with 1 to 2 cups sugar for each quart of juice, depending on tartness of fruit. Pour into containers, leaving head space (p. 7). Seal and freeze.

Raspberries

● **Whole.** Raspberries may be frozen in sugar or sirup or unsweetened. Seedy berries are best for use in making purees or juice.

Select fully ripe, juicy berries. Sort, wash carefully in cold water, and drain thoroughly.

Sugar pack. To 1 quart (1⅓ pounds) berries add ¾ cup sugar and mix carefully to avoid crushing. Put into containers, leaving head space (p. 7). Seal and freeze.

Sirup pack. Put berries into containers and cover with cold 40-percent sirup (p. 10), leaving head space (p. 7). Seal and freeze.

Unsweetened pack. Put berries into containers, leaving head space (p. 7). Seal and freeze.

● **Crushed or puree.** Prepare as for whole raspberries; then crush or press through a sieve for puree.

To 1 quart (2 pounds) crushed berries or puree add ¾ to 1 cup sugar, depending on sweetness of fruit. Mix until sugar is dissolved. Put into containers, leaving head space (p. 7). Seal and freeze.

● **Juice.** For beverage use, select as for whole raspberries. For jelly making, select as recommended in specific jelly recipe. Crush and heat berries slightly to start flow of juice. Strain in a jelly bag to extract juice.

Sweeten with ½ to 1 cup sugar for each quart of juice if desired. Pour into containers, leaving head space (p. 7). Seal and freeze.

Rhubarb

- **Stalks or pieces.** Choose firm, tender, well-colored stalks with good flavor and few fibers. Wash, trim, and cut into 1- or 2-inch pieces or in lengths to fit the package. Heating rhubarb in boiling water for 1 minute and cooling promptly in cold water helps retain color and flavor.

  Unsweetened pack. Pack either raw or preheated rhubarb tightly into containers without sugar. Leave head space (p. 7). Seal and freeze.

  Sirup pack. Pack either raw or preheated rhubarb tightly into containers, cover with cold 40-percent sirup (p. 10). Leave head space (p. 7). Seal and freeze.

- **Puree.** Prepare rhubarb as for rhubarb stalks or pieces. Add 1 cup water to 1½ quarts (2 pounds) rhubarb and boil 2 minutes. Cool and press through a sieve. With 1 quart (2 pounds) puree mix ⅔ cup sugar. Pack into containers, leaving head space (p. 7). Seal and freeze.

- **Juice.** Select as for rhubarb stalks or pieces. Wash, trim, and cut into pieces 4 to 6 inches long. Add 1 quart water to 4 quarts (5 pounds) rhubarb and bring just to a boil. Press hot fruit in jelly bag to extract juice. Cool. Sweeten, if desired, using ½ cup sugar to a quart of juice. Pour into containers, leaving head space (p. 7). Seal and freeze.

Strawberries

- **Whole.** Choose firm, ripe, red berries preferably with a slightly tart flavor. Large berries are better sliced or crushed. Sort berries, wash them in cold water, drain well, and remove hulls.

  Sugar and sirup packs make better quality frozen strawberries than berries packed without sweetening.

  Sirup pack. Put berries into containers and cover with cold 50-percent sirup (p. 10), leaving head space (p. 7). Seal and freeze.

  Sugar pack. Add ¾ cup sugar to 1 quart (1⅓ pounds) strawberries and mix thoroughly. Put into containers, leaving head space (p. 7). Seal and freeze.

  Unsweetened pack. Pack into containers, leaving head space (p. 7). For better color, cover with water containing 1 teaspoon crystalline ascorbic acid to each quart of water. Seal and freeze.

- **Sliced or crushed.** Prepare for packing as for whole strawberries; then slice, or crush partially or completely.

  To 1 quart (1½ pounds) berries add ¾ cup sugar; mix thoroughly. Pack into containers, leaving head space (p. 7). Seal and freeze.

- **Puree.** Prepare strawberries as for freezing whole. Then press berries through a sieve. To 1 quart (2 pounds) puree add ⅔ cup sugar and mix well. Put into containers, leaving head space (p. 7). Seal and freeze.

- **Juice.** Choose fully ripe berries. Sort and wash them in cold water. Drain well and remove hulls. Crush berries and strain juice through a jelly bag. Sweeten with ⅔ to 1 cup sugar to each quart of juice, or leave unsweetened. Pour into containers, leaving head space (p. 7). Seal and freeze.

Points on freezing vegetables

Best for freezing are fresh, tender vegetables right from the garden. The fresher the vegetables when frozen the more satisfactory will be your product.

First steps

Washing is the first step in the preparation of most vegetables for freezing. However, lima beans, green peas, and other vegetables that are protected by pods may not need to be washed.

Wash vegetables thoroughly in cold water. Lift them out of the water as grit settles to the bottom of the pan.

Sort vegetables according to size for heating and packing unless they are to be cut into pieces of uniform size.

Peel, trim, and cut into pieces, as directed for each vegetable on pages 36 to 41.

Heating before packing

An important step in preparing vegetables for freezing is heating or "blanching" before packing. Practically every vegetable, except green pepper, maintains better quality in frozen storage if heated before packing.

The reason for heating vegetables before freezing is that it slows or stops the action of enzymes. Up until the time vegetables are ready to pick, enzymes help them grow and mature. After that they cause loss of flavor and color. If vegetables are not heated enough the enzymes continue to be active during frozen storage. Then the vegetables may develop off-flavors, discolor, or toughen so that they may be unappetizing in a few weeks.

Heating also wilts or softens vegetables and makes them easier to pack. Heating time varies with the vegetable and size of pieces.

To heat in boiling water. For home freezing, the most satisfactory way to heat practically all vegetables is in boiling water. Use a blancher, which has a blanching basket and cover. Or fit a wire basket into a large kettle, and add the cover.

For each pound of prepared vegetable use at least 1 gallon of boiling water in the blancher or kettle. Put vegetables in blanching basket or wire basket and lower into the boiling water. A wire cover for the basket can be used to keep the vegetables down in the boiling water.

Put lid on blancher or kettle and start counting time immediately. Keep heat high for time given in directions for vegetable you are freezing. Heat 1 minute longer than the time specified if you live 5,000 feet or more above sea level.

To heat in steam. In this publication, heating in steam is recommended for a few vegetables. For broccoli, pumpkin, sweetpotatoes, and winter squash both steaming and boiling are satisfactory methods.

To steam, use a kettle with a tight lid and a rack that holds a steaming basket at least 3 inches above the bottom of the kettle. Put an inch or two of water in the kettle and bring the water to a boil.

Put vegetables in the basket in a single layer so that steam reaches all parts quickly. Cover the kettle and keep heat high. Start counting steaming time as soon as the lid is on. Steam 1 minute longer than the time specified in directions if you live 5,000 feet or more above sea level.

Other ways to heat. Pumpkin, sweetpotatoes, and winter squash may be heated in a pressure cooker or in the oven before freezing. Mushrooms may be heated in fat in a fry pan. Tomatoes for juice may be simmered.

Cooling

After vegetables are heated they should be cooled quickly and thoroughly to stop the cooking.

To cool vegetables heated in boiling water or steam, plunge the basket of vegetables immediately into a large quantity of cold water—60° F. or below. Change water frequently or use cold running water or iced water. If ice is used, you'll need about 1 pound of ice for each pound of vegetable. It will take about as long to cool the food as it does to heat it. When the vegetable is cool, remove it from the water and drain thoroughly.

To cool vegetables heated in the oven, a pressure cooker, or a fry pan—set pan of food in water and change water to speed cooling.

Dry pack more practical

Either dry or brine pack may be used for most vegetables to be frozen. However, in this publication the dry pack is recommended for all vegetables, because preparation for freezing and serving is easier.

Table of vegetable yields

The table on page 31 will help you figure the amount of frozen food you can get from a given amount of a fresh vegetable. The number of pints of frozen vegetables you get depends on the quality, condition, maturity, and variety—and on the way the vegetable is trimmed and cut.

Approximate yield of frozen vegetables from fresh

VEGETABLE	FRESH, AS PURCHASED OR PICKED	FROZEN
Asparagus	1 crate (12 2-lb. bunches) 1 to 1½ lb.	15 to 22 pt. 1 pt.
Beans, lima (in pods)	1 bu. (32 lb.) 2 to 2½ lb.	12 to 16 pt. 1 pt.
Beans, snap, green, and wax	1 bu. (30 lb.) ⅔ to 1 lb.	30 to 45 pt. 1 pt.
Beet greens	15 lb. 1 to 1½ lb.	10 to 15 pt. 1 pt.
Beets (without tops)	1 bu. (52 lb.) 1¼ to 1½ lb.	35 to 42 pt. 1 pt.
Broccoli	1 crate (25 lb.) 1 lb.	24 pt. 1 pt.
Brussels sprouts	4 quart boxes 1 lb.	6 pt. 1 pt.
Carrots (without tops)	1 bu. (50 lb.) 1¼ to 1½ lb.	32 to 40 pt. 1 pt.
Cauliflower	2 medium heads 1⅓ lb.	3 pt. 1 pt.
Chard	1 bu. (12 lb.) 1 to 1½ lb.	8 to 12 pt. 1 pt.
Collards	1 bu. (12 lb.) 1 to 1½ lb.	8 to 12 pt. 1 pt.
Corn, sweet (in husks)	1 bu. (35 lb.) 2 to 2½ lb.	14 to 17 pt. 1 pt.
Kale	1 bu. (18 lb.) 1 to 1½ lb.	12 to 18 pt. 1 pt.
Mustard greens	1 bu. (12 lb.) 1 to 1½ lb.	8 to 12 pt. 1 pt.
Peas	1 bu. (30 lb.) 2 to 2½ lb.	12 to 15 pt. 1 pt.
Peppers, sweet	⅔ lb. (3 peppers)	1 pt.
Pumpkin	3 lb.	2 pt.
Spinach	1 bu. (18 lb.) 1 to 1½ lb.	12 to 18 pt. 1 pt.
Squash, summer	1 bu. (40 lb.) 1 to 1¼ lb.	32 to 40 pt. 1 pt.
Squash, winter	3 lb.	2 pt.
Sweetpotatoes	⅔ lb.	1 pt.

Home Freezing of Fruits and Vegetables (31)

Freezing snap beans

Other vegetables may be frozen in much the same way as snap beans. Beans are heated in boiling water before they are frozen—the most satisfactory home method for nearly all vegetables.

● Select young, tender, stringless beans that snap when broken. Allow ⅔ to 1 pound of fresh beans for 1 pint frozen. Wash thoroughly.

278A

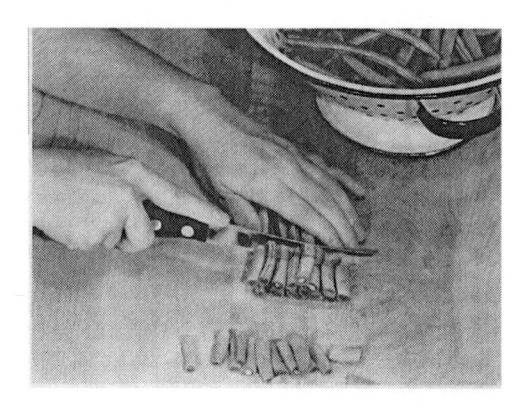

● Cut beans into 1- or 2-inch pieces or slice them lengthwise.

280A

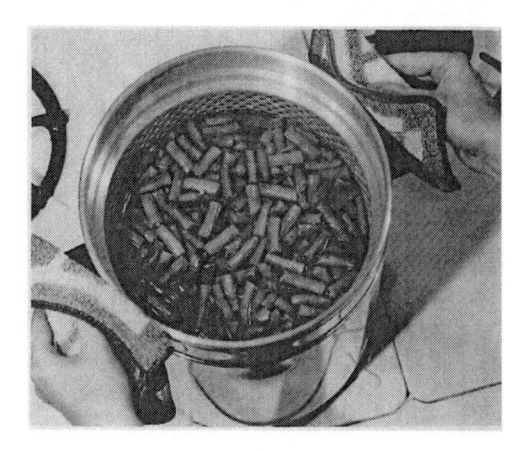

● Put beans in blanching basket, lower basket into boiling water, and cover. Heat for 3 minutes. Keep heat high under the water.

281A

152 (32) **Home Freezing of Fruits and Vegetables**

● Plunge basket of heated beans into cold water to stop the cooking. It takes about as long to cool vegetables as to heat them. When beans are cool, remove them from water and drain.

282A

● Pack the beans into bags or other containers. A stand to hold the bags makes filling easier. A funnel helps keep the sealing edges clean.

283A

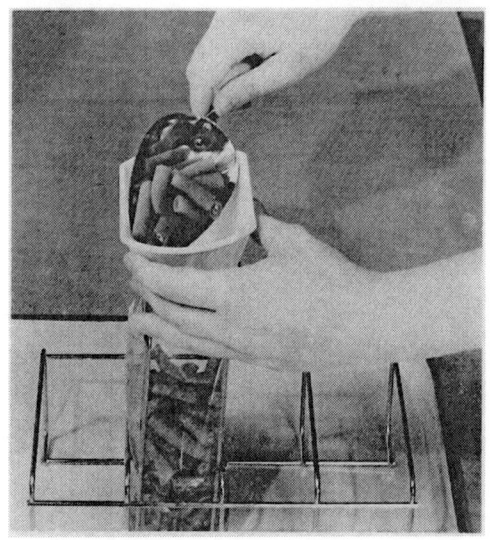

● Leave ½-inch head space and seal by twisting and folding back top of bag and tying with a string. Freeze beans at once. Store at 0° F. or below. If the bags used are of materials that become brittle at low temperatures, they need an outside carton for protection.

284A

Home Freezing of Fruits and Vegetables (33) 153

Freezing broccoli

Broccoli, like all vegetables, is best frozen as soon as possible after it is picked. Allow about 1 pound fresh broccoli for each pint frozen. Because broccoli packs loosely, no head space need be allowed. Containers other than those shown here also may be used for packing.

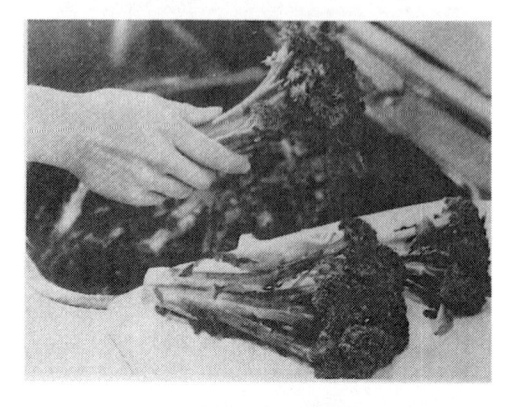

- Select tight, compact, dark-green heads with tender stalks free from woodiness. Trim off large leaves and tough parts of stems and wash thoroughly.

 If necessary, soak stalks for ½ hour in salt water (made of 4 teaspoons salt to each gallon of water) to remove insects.

- Cut broccoli lengthwise into uniform pieces, leaving heads about 1½ inches across—to insure uniform heating and make attractive pieces for serving.

- Steam pieces by placing them in blanching basket over rapidly boiling water. Cover kettle, keep heat high, and steam for 5 minutes. Or heat pieces in boiling water 3 minutes, as is shown for snap beans (p. 32).

- Remove basket from boiling water. Cool broccoli by plunging basket into cold water.

- Lift basket from cold water as soon as broccoli is cool and let drain a few minutes.

- Pack broccoli so some heads are at each end of the container—to get more broccoli in the package. No head space is needed. Press lid on firmly to seal. Freeze at once. Store at 0° F. or below.

Home Freezing of Fruits and Vegetables (35) 155

Directions for vegetables

Asparagus

Select young, tender stalks with compact tips. Sort according to thickness of stalk.

Wash asparagus thoroughly and cut or break off and discard tough parts of stalks. Leave spears in lengths to fit the package or cut in 2-inch lengths.

Heat stalks in boiling water according to thickness of stalk:

Small stalks.........	2 minutes
Medium stalks.........	3 minutes
Large stalks...........	4 minutes

Cool promptly in cold water and drain.

Pack into containers, leaving no head space. When packing spears, alternate tips and stem ends. In containers that are wider at the top than bottom, pack asparagus with tips down. Seal and freeze.

Beans, lima

Select well-filled pods. Beans should be green but not starchy or mealy. Shell and sort according to size, or leave beans in pods to be shelled after heating and cooling. Heat in boiling water:

Small beans or pods......	2 minutes
Medium beans or pods .	3 minutes
Large beans or pods......	4 minutes

Cool promptly in cold water and drain.

Pack into containers, leaving ½-inch head space. Seal and freeze.

Beans, shell, green

Select pods that are plump, not dry or wrinkled. Shell the beans. Heat in boiling water 1 minute. Cool promptly in cold water and drain.

Pack into containers, leaving ½-inch head space. Seal and freeze.

Beans, snap, green, or wax

Select young, tender, stringless beans that snap when broken. Wash thoroughly; then remove ends.

Cut in 1- or 2-inch pieces, or slice lengthwise into strips for frenched (julienne-style) snap beans.

Heat in boiling water for 3 minutes. Chill promptly in cold water and drain.

Pack into containers, leaving ½-inch head space. Seal and freeze.

Beans, soybeans, green

Select firm, well-filled, bright-green pods. Wash. Heat beans in pods 5 minutes in boiling water, and cool promptly in cold water. Squeeze soybeans out of pods.

Pack soybeans into containers, leaving ½-inch head space. Seal and freeze.

Beets

Select young or mature beets not more than 3 inches across.

Wash and sort according to size. Trim tops, leaving ½ inch of stems.

Cook in boiling water until tender—for small beets, 25 to 30 minutes; for medium-size beets, 45 to 50 minutes. Cool promptly in cold water. Peel and cut into slices or cubes.

Pack beets into containers, leaving ½-inch head space. Seal and freeze.

Broccoli

Select tight, compact, dark-green heads with tender stalks free from woodiness. Wash, peel stalks, and trim. If necessary to remove insects, soak for ½ hour in a solution made of 4 teaspoons salt to 1 gallon cold water. Split lengthwise into pieces so that flowerets are not more than 1½ inches across.

Heat in steam 5 minutes or in boiling water 3 minutes. Cool promptly in cold water and drain.

Pack broccoli into containers, leaving no head space. Seal and freeze.

Brussels sprouts

Select green, firm, and compact heads. Examine heads carefully to make sure they are free from insects. Trim, removing coarse outer leaves. Wash thoroughly. Sort into small, medium, and large sizes.

Heat in boiling water:

Small heads	3 minutes
Medium heads	4 minutes
Large heads	5 minutes

Cool promptly in cold water and drain.

Pack brussels sprouts into containers, leaving no head space. Seal and freeze.

Cabbage or chinese cabbage

Frozen cabbage or chinese cabbage is suitable for use only as a cooked vegetable.

Select freshly picked, solid heads. Trim coarse outer leaves from head. Cut into medium to coarse shreds or thin wedges, or separate head into leaves. Heat in boiling water 1½ minutes.

Cool promptly in cold water and drain.

Pack cabbage into containers, leaving ½-inch head space. Seal and freeze.

Carrots

Select tender, mild-flavored carrots. Remove tops, wash, and peel. Leave small carrots whole. Cut others into ¼-inch cubes, thin slices, or lengthwise strips.

Heat in boiling water:

Whole carrots, small	5 minutes
Diced or sliced	2 minutes
Lengthwise strips	2 minutes

Cool promptly in cold water and drain.

Pack carrots into containers, leaving ½-inch head space. Seal and freeze.

Cauliflower

Choose firm, tender, snow white heads. Break or cut into pieces about 1 inch across. Wash well. If necessary to remove insects, soak for 30 minutes in a solution of salt and water—4 teaspoons salt to each gallon of water. Drain.

(Continued on page 38)

Cauliflower—continued

Heat in boiling water containing 4 teaspoons salt to a gallon for 3 minutes. Cool promptly in cold water and drain.

Pack cauliflower into containers, leaving no head space. Seal and freeze.

Celery

Select crisp, tender stalks, free from coarse strings and pithiness.

Wash thoroughly, trim, and cut stalks into 1-inch lengths.

Heat for 3 minutes in boiling water. Cool promptly in cold water and drain.

Pack celery into containers, leaving $\frac{1}{2}$-inch head space. Seal and freeze.

Corn, sweet

● **Whole-kernel and cream-style.** Select ears with plump, tender kernels and thin, sweet milk. If the milk is thick and starchy it is better to freeze corn as cream-style.

Husk ears, remove silk, and wash the corn. Heat ears in boiling water for 4 minutes. Cool promptly in cold water and drain.

For whole-kernel corn, cut kernels from cob at about two-thirds the depth of the kernels.

For cream-style corn, cut corn from the cob at about the center of the kernels. Scrape the cobs with the back of the knife to remove the juice and the heart of the kernel.

Pack corn into containers, leaving $\frac{1}{2}$-inch head space. Seal and freeze.

● **On-the-cob.** Select same as for whole-kernel sweet corn.

Husk, remove silk, wash, and sort ears according to size.

Heat in boiling water:

Small ears.................. 7 minutes
($1\frac{1}{4}$ inches or less in diameter)
Medium ears............... 9 minutes
($1\frac{1}{4}$ to $1\frac{1}{2}$ inches in diameter)
Large ears.................. 11 minutes
(over $1\frac{1}{2}$ inches in diameter)

Cool promptly in cold water and drain.

Pack ears into containers or wrap in moisture-vapor-resistant material. Seal and freeze.

Greens—beet greens, chard, collards, kale, mustard greens, spinach, turnip greens

Select young, tender leaves. Wash well. Remove tough stems and imperfect leaves. Cut leaves of chard into pieces as desired.

Heat in boiling water for the following periods:

Beet greens, kale, chard, mustard greens, turnip greens.   2 minutes
Collards.... .... 3 minutes
Spinach and New Zealand spinach.   2 minutes
Very tender leaves....$1\frac{1}{2}$ minutes

Cool promptly in cold water and drain.

Pack greens into containers, leaving $\frac{1}{2}$-inch head space. Seal and freeze.

Kohlrabi

Select young, tender, mild-flavored kohlrabi, small to medium in size. Cut off tops and roots. Wash, peel, and leave whole or dice in 1/2-inch cubes.

Heat in boiling water:

Whole kohlrabi............ 3 minutes
Cubes..... .... 1 minute

Cool promptly in cold water and drain.

Pack whole kohlrabi into containers or wrap in moisture-vapor-resistant material. Seal and freeze.

Pack cubes into containers, leaving 1/2-inch head space. Seal and freeze.

Mushrooms

Choose mushrooms free from spots and decay. Sort according to size. Wash thoroughly in cold water. Trim off ends of stems. If mushrooms are larger than 1 inch across, slice them or cut them into quarters.

Mushrooms may be steamed or heated in fat in a fry pan.

● **To steam.** Mushrooms to be steamed have better color if given antidarkening treatment first.

Dip for 5 minutes in a solution containing 1 teaspoon lemon juice or 1 1/2 teaspoons citric acid to a pint of water. Then steam:

Whole mushrooms....... 5 minutes
 (not larger than 1 inch
 across)
Buttons or quarters... 3 1/2 minutes
Slices. .. .. 3 minutes

Cool promptly in cold water and drain.

● **To heat in fry pan.** Heat small quantities of mushrooms in table fat in an open fry pan until almost done.

Cool in air or set pan in which mushrooms were cooked in cold water.

Pack into containers, leaving 1/2-inch head space. Seal and freeze.

Okra

Select young, tender, green pods. Wash thoroughly. Cut off stems in such a way as not to cut open seed cells.

Heat in boiling water:

Small pods..... .... 3 minutes
Large pods . ..... 4 minutes

Cool promptly in cold water and drain.

Leave whole or slice crosswise.

Pack into containers, leaving 1/2-inch head space. Seal and freeze.

Parsnips

Choose small to medium-size parsnips that are tender and free from woodiness. Remove tops, wash, peel, and cut in 1/2-inch cubes or slices.

Heat in boiling water 2 minutes. Cool promptly in cold water; drain.

Pack into containers, leaving 1/2-inch head space. Seal and freeze.

Peas, field (blackeye)

Select well-filled flexible pods with tender seeds. Shell peas, discarding those that are hard.

Heat in boiling water for 2 minutes. Cool promptly in cold water and drain.

Pack into containers, leaving 1/2-inch head space. Seal and freeze.

Home Freezing of Fruits and Vegetables (39)

Peas, green

Choose bright-green, plump, firm pods with sweet, tender peas. Do not use immature or tough peas.

Shell peas. Heat in boiling water 1½ minutes. Cool promptly in cold water and drain.

Pack peas into containers, leaving ½-inch head space. Seal and freeze.

Peppers, sweet or hot

● **Sweet.** Peppers frozen without heating are best for use in uncooked foods. Heated peppers are easier to pack and good for use in cooking.

Select firm, crisp, thick-walled peppers. Wash, cut out stems, cut in half, and remove seeds. If desired, cut into ½-inch strips or rings.

Heat in boiling water if desired:

Halves . . . . 3 minutes
Slices . . . . 2 minutes

Cool promptly in cold water and drain.

If peppers have not been heated, pack into containers, leaving no head space. Seal and freeze. If peppers have been heated, leave ½-inch head space.

● **Hot.** Wash and stem peppers Pack into small containers, leaving no head space. Seal and freeze.

Pimientos

Select firm, crisp, thick-walled pimientos.

To peel, first roast pimientos in an oven at 400° F. (hot oven) for 3 to 4 minutes. Remove charred skins by rinsing pimientos in cold water. Drain.

Pack pimientos into containers leaving ½-inch head space. Seal and freeze.

Pumpkin

Select full-colored, mature pumpkin with texture that is fine rather than coarse and stringy.

Wash, cut into quarters or smaller pieces, and remove seeds. Cook pumpkin pieces until soft in boiling water, in steam, in a pressure cooker, or in the oven.

Remove pulp from rind and mash it or press it through a sieve.

To cool, place pan containing pumpkin in cold water. Stir pumpkin occasionally.

Pack into containers, leaving ½-inch head space. Seal and freeze.

Rutabagas

Select young, tender, medium-size rutabagas with no tough fibers. Cut off tops, wash, and peel.

● **Cubed.** Cut into ½-inch cubes. Heat in boiling water for 2 minutes. Cool promptly in cold water; drain. Pack into containers, leaving ½-inch head space. Seal and freeze.

● **Mashed.** Cut rutabagas in pieces. Cook until tender in boiling water and drain. Mash or press through a sieve.

To cool, place pan containing rutabagas in cold water. Stir rutabagas occasionally. Pack into containers, leaving ½-inch head space. Seal and freeze.

Squash, summer or winter

● **Summer.** Select young squash with small seeds and tender rind. Wash, cut in ½-inch slices. Heat in boiling water for 3 minutes. Cool squash promptly in cold water and drain.

Pack into containers, leaving ½-inch head space. Seal and freeze.

● **Winter.** Select firm, mature squash. Wash, cut into pieces, and remove seeds. Cook pieces until soft in boiling water, in steam, in a pressure cooker, or in the oven. Remove pulp from rind and mash or press through a sieve.

To cool, place pan containing squash in cold water and stir squash occasionally.

Pack into containers, leaving ½-inch head space. Seal and freeze.

Sweetpotatoes

Sweetpotatoes may be packed whole, sliced, or mashed.

Choose medium to large mature sweetpotatoes that have been cured. Sort according to size, and wash.

Cook until almost tender in water, in steam, in a pressure cooker, or in the oven. Let stand at room temperature until cool. Peel sweetpotatoes; cut in halves, slice, or mash.

If desired, to prevent darkening, dip whole sweetpotatoes or slices for 5 seconds in a solution of 1 tablespoon citric acid or ½ cup lemon juice to 1 quart water.

To keep mashed sweetpotatoes from darkening, mix 2 tablespoons orange or lemon juice with each quart of mashed sweetpotatoes.

Pack into containers, leaving ½-inch head space. Seal and freeze.

● **For variety.** Roll cooked sweetpotato slices in sugar. Pack into containers, leaving ½-inch head space. Seal and freeze.

Or pack whole or sliced cooked sweetpotatoes in containers, cover with cold sirup (made of equal parts by measure of sugar and water). Leave head space (see p. 7, for liquid pack). Seal and freeze.

Tomatoes

● **Juice.** Wash, sort, and trim firm, vine-ripened tomatoes. Cut in quarters or eighths. Simmer 5 to 10 minutes. Press through a sieve. If desired, season with 1 teaspoon salt to each quart of juice. Pour into containers, leaving head space (see p. 7, for juices). Seal and freeze.

● **Stewed.** Remove stem ends, peel, and quarter ripe tomatoes. Cover and cook until tender (10 to 20 minutes). Place pan containing tomatoes in cold water to cool. Pack into containers, leaving head space (see p. 7, for liquid pack). Seal and freeze.

Turnips

Select small to medium, firm turnips that are tender and have a mild flavor. Wash, peel, and cut into ½-inch cubes. Heat in boiling water for 2 minutes. Cool promptly in cold water and drain.

Pack into containers, leaving ½-inch head space. Seal and freeze.

How to use frozen fruits and vegetables

Fruits

Serving uncooked. Frozen fruits need only to be thawed, if they are to be served raw.

For best color and flavor, leave fruit in the sealed container to thaw. Serve as soon as thawed; a few ice crystals in the fruit improve the texture for eating raw.

Frozen fruit in the package may be thawed in the refrigerator, at room temperature, or in a pan of cool water. Turn package several times for more even thawing.

Allow 6 to 8 hours on a refrigerator shelf for thawing a 1-pound package of fruit packed in sirup. Allow 2 to 4 hours for thawing a package of the same size at room temperature—½ to 1 hour for thawing in a pan of cool water.

Fruit packed with dry sugar thaws slightly faster than that packed in sirup. Both sugar and sirup packs thaw faster than unsweetened packs.

Thaw only as much as you need at one time. If you have leftover thawed fruit it will keep better if you cook it. Cooked fruit will keep in the refrigerator for a few days.

Cooking. First thaw fruits until pieces can be loosened. Then cook as you would cook fresh fruit. If there is not enough juice to prevent scorching, add water as needed. If the recipe calls for sugar, allow for any sweetening that was added before freezing.

Frozen fruits often have more juice than called for in recipes for baked products using fresh fruits. In that case use only part of the juice, or add more thickening for the extra juice.

Using crushed fruit and purees. Serve crushed fruit as raw fruit—after it is partially or completely thawed. Or use it after thawing as a topping for ice cream or cakes, as a filling for sweet rolls, or for jam.

Use thawed purees in puddings, ice cream, sherbets, jams, pies, ripple cakes, fruit-filled coffee cake, and rolls.

Serving juice. Serve frozen fruit juice as a beverage—after it is thawed but while it is still cold. Some juices, such as sour cherry, plum, grape, and berry juices, may be diluted $\frac{1}{3}$ to $\frac{1}{2}$ with water or a bland juice.

Vegetables

The secret of cooking frozen vegetables successfully is to cook the vegetable until just tender. That way you save vitamins, bright color, and fresh flavor.

Frozen vegetables may be cooked in a small amount of water or in a pressure saucepan, or by baking or panfrying.

Cooking in a small amount of water. You should cook most frozen vegetables without thawing them first. Leafy vegetables, such as spinach, cook more evenly if thawed just enough to separate the leaves before cooking. Corn-on-the-cob should be partially thawed before cooking, so that the cob will be heated through by the time the corn is cooked. Holding corn after thawing or cooking causes sogginess.

Bring water to a boil in a covered saucepan. The amount of water to use depends on the vegetable and the size of the package. For most vegetables one-half cup of water is enough for a pint package. The frost in the packages furnishes some additional moisture.

Put the frozen vegetable in the boiling water, cover the pan, and bring the water quickly back to a boil. To insure uniform cooking, it may be necessary to separate pieces carefully with a fork. When the water is boiling throughout the pan, reduce the heat and start counting time. Be sure pan is covered to keep in the steam, which aids in cooking. Cook gently until vegetables are just tender.

Add seasonings as desired and serve immediately.

The following timetable shows about how long it takes to cook tender one pint of various frozen vegetables—and how much water to use. Use the table only as a general guide. Cooking times vary among varieties and with the maturity of the vegetable when it is frozen.

The time required for cooking vegetables is slightly longer at high than at low altitudes because the temperature of boiling water decreases about 2° F. with each 1,000 feet above sea level.

Timetable for cooking frozen vegetables in a small amount of water [1]

VEGETABLE	Time to allow after water returns to boil [2] Minutes	VEGETABLE	Time to allow after water returns to boil [2] Minutes
Asparagus	5–10	Chard	8–10
Beans, lima:		Corn:	
Large type	6–10	Whole-kernel	3–5
Baby type	15–20	On-the-cob	3–4
Beans, snap, green, or wax:		Kale	8–12
1-inch pieces	12–18	Kohlrabi	8–10
Julienne	5–10	Mustard greens	8–15
Beans, soybeans, green	10–20	Peas, green	5–10
Beet greens	6–12	Spinach	4–6
Broccoli	5–8	Squash, summer	10–12
Brussels sprouts	4–9	Turnip greens	15–20
Carrots	5–10	Turnips	8–12
Cauliflower	5–8		

[1] Use ½ cup of lightly salted water for each pint of vegetable with these exceptions: Lima beans, 1 cup; corn-on-the-cob, water to cover.
[2] Time required at sea level; slightly longer time is required at higher altitudes.

Cooking in a pressure saucepan. Follow directions and cooking times specified by the manufacturer of your saucepan.

Baking. Many frozen vegetables may be baked in a covered casserole. Partially defrost vegetable to separate pieces.

Put vegetable in a greased casserole; add seasonings as desired. Cover and bake until just tender.

The time it takes to bake a vegetable varies with size of pieces and how much you thaw them before baking.

Approximate time for baking most thawed vegetables is 45 minutes at 350° F. (moderate oven). Slightly more time may be required if other foods are being baked at the same time.

To bake corn-on-the-cob, partially thaw the ears first. Brush with melted butter or margarine, salt, and roast at 400° F. (hot oven) about 20 minutes.

Pan frying. Use a heavy fry pan with cover. Place about 1 tablespoon fat in pan. Add 1 pint frozen vegetable, which has been thawed enough to separate pieces. Cook covered over moderate heat. Stir occasionally. Cook until just tender. Season to taste, and serve immediately.

Peas, asparagus, and broccoli will cook tender in a fry pan in about 10 minutes. Mushrooms will be done in 10 to 15 minutes and snap beans in 15 to 20 minutes.

Other ways to prepare frozen vegetables. Vegetables that are cooked until tender before freezing need only to be seasoned and heated before serving. Cooked frozen vegetables can be used in many dishes in the same ways as cooked fresh vegetables. They may be creamed or scalloped, served au gratin, or added to souffles, cream soups, or salads.

Pumpkin, winter squash, and sweetpotatoes may be thawed and used as the main ingredient in pie fillings.

Index

	Page
Altitude, effect of—	
on cooking time for frozen vegetables	43–44
on heating time for vegetables	29–30
Apples	18
Applesauce	18
Apricots	18–19
Ascorbic acid, to keep fruit from darkening	11
Ascorbic acid mixtures, to keep fruit from darkening	12
Asparagus	36
Avocados	19
Beans—	
lima	36
shell, green	36
snap, green	32–33, 36
soybeans, green	36
wax	36
Beet greens	38
Beets	36–37
Berries—	
blackberries	19–20
blueberries	20
boysenberries	19–20
dewberries	19–20
elderberries	20
gooseberries	23
huckleberries	20
loganberries	19–20
raspberries	27
strawberries	14–15, 28
youngberries	19–20
Blackberries	19–20
Blueberries	20
Boysenberries	19–20
Brine pack	30
Broccoli	34–35, 37
Brussels sprouts	37
Cabbage	37
Cantaloup	24–25
Carrots	37
Cauliflower	37–38
Celery	38
Chard	38
Cherries—	
sour	20–21
sweet	21–22
sweet with sour	21–22

	Page
Chinese cabbage	37
Citric acid, to keep fruits from darkening	12
Coconut	22
Collards	38
Containers—	
care	6
cost	6
kind	4
reuse	6
sealing	5
shape	5
size	5
Cooking—	
frozen fruits	42
frozen vegetables	43–45
Cooling vegetables	30
Corn, sweet—	
cream-style	38
on-the-cob	38
whole-kernel	38
Cranberries	22
Crenshaw melons	24
Crushed fruit	9
(See also directions for each fruit.)	
Currants	22
Darkening, agents to retard—	
ascorbic acid	11
ascorbic acid mixtures	12
citric acid	12
lemon juice	12
steam	12
Dates	23
Dewberries	19–20
Dry ice, to prevent thawing	8
Dry pack	30
Elderberries	20
Enzymes	29
Equipment	4–6, 9, 29–30
Figs	23
Freezer—	
costs of owning and operating	3
failure to operate	8
loading the freezer	7
Freezing—	
accessories	6
amount to freeze at one time	7
temperature	7
Fruit cocktail	23

166 (46) Home Freezing of Fruits and Vegetables

	Page
Fruits—	
darkening, to keep from	11
frozen, use—	
crushed fruit and purees	43
for cooking	42
for serving raw	42
juice	43
thawing	42
packing in sirup, with sugar, or unsweetened	10–11
points on freezing	9
preparation for freezing	9
selection	3–4, 9
(See also directions for each fruit.)	
sirup pack	10–11
sirup proportions	10
sugar pack	10, 11
thawing	42
unsweetened pack	10, 11
varieties suitable for freezing	3
yield of frozen fruit from fresh	12–13
Gooseberries	23
Grapefruit	23–24
Grapes	24
Greens—	
beet greens	38
chard	38
collards	38
kale	38
mustard greens	38
New Zealand spinach	38
salad greens (not suitable for freezing)	4
spinach	38
turnip greens	38
Head space, amount to allow	6, 7
Heating vegetables before packing—	
in boiling water	29
in steam	30
other ways	30
Honeydew melons	24–25
Huckleberries	20
Inventory of frozen foods	8
Kale	38
Kohlrabi	39
Lemon juice, to keep fruit from darkening	12
Lettuce (not suitable for freezing)	4
Loganberries	19–20
Melons—	
cantaloup	24
crenshaw	24
honeydew	24
Persian	24
watermelon	24
Mushrooms	39
Mustard greens	38
Nectarines	25

	Page
New Zealand spinach	38
Okra	39
Onions, green (not suitable for freezing)	4
Oranges	23–24
Packing fruits—	
in sirup	10–11
unsweetened	10, 11
with sugar	10, 11
Packing vegetables—	
brine pack	30
dry pack	30
Parsnips	39
Peaches	16–17, 25
Pears	26
Peas—	
field (blackeye)	39
green	40
Peppers—	
hot	40
sweet	40
Persian melons	24–25
Persimmons	26
Pimientos	40
Pineapple	26
Plums	26–27
Preparation for freezing—	
fruits	9
vegetables	29–30
Preparation for the table—	
fruits	42
vegetables	43–45
Pressure saucepan for cooking frozen vegetables	44
Prunes	26–27
Pumpkin	40
Puree	9
(See also directions for each fruit.)	
Radishes (not suitable for freezing)	4
Raspberries	27
Refreezing	8
Rhubarb	28
Rutabagas	40
Sirup pack for fruits	10
Spinach	38
Spoilage, prevention in emergency	8
Squash—	
summer	41
winter	41
Steaming—	
to heat vegetables before packing	30
to keep fruit from darkening	12

Home Freezing of Fruits and Vegetables (47) 167

	Page
Storage—	
temperature	8
time	8
Strawberries	14–15, 28
Sugar pack for fruits	11, 12
Sweetpotatoes	41
Thawing fruits	42
Timetable for cooking vegetables in small amount of water	44
Tomatoes—	
juice	41
stewed	41
Tomatoes (not suitable for freezing)	4
Turnip greens	38
Turnips	41
Unsweetened pack for fruits	11
Use of frozen fruits	42–43
Use of frozen vegetables	43–45
Varieties suitable for freezing	3

	Page
Vegetables—	
cooking	43–45
cooling	30
dry pack	30
frozen, use—	
baking	44
cooking in pressure saucepan	44
cooking in small amount of water	43–44
pan frying	44
timetable for cooking	44
heating before packing—	
in boiling water	29
in steam	30
other ways	30
points on freezing	29
preparation for freezing	29–30
selection	3–4, 29
(See also directions for each vegetable.)	
timetable for cooking—	
in small amount of water	44
varieties suitable for freezing	3
yield of frozen from fresh	30–31
Watermelon	24–25
Youngberries	19–20

HOME FREEZING OF POULTRY

Home Freezing of Poultry

You can freeze poultry simply and easily at home. There is no better way to preserve it. Frozen poultry is similar to unfrozen in flavor, texture, color, and nutritive value.

Poultry can go into the freezer in a variety of forms. Uncooked poultry may be frozen whole, in halves, or in serving-size pieces. Cooked poultry—pieces with bone in, large or small pieces removed from the bone, and slices—may be frozen with or without gravy or broth. Cooked poultry meat may also be frozen in sandwiches or in various combination main dishes.

SELECTING POULTRY TO FREEZE

Select for freezing only fresh, high-quality poultry that is of suitable age and size for the intended use.

Choose well-fleshed birds with fat well distributed. If you buy ready-to-cook poultry, look for birds with few skin blemishes.

Fresh.—Fresh-killed poultry is best for home freezing.

If you buy whole birds displayed in film or cut-up birds in tray packs, do not plan to freeze the birds in these wrappings. Such wrappings are not designed for freezer storage.

It may be more economical sometimes to buy poultry that is already frozen in packaging designed for freezer storage.

Suitable for intended use.—Have in mind the intended use of the frozen product and choose poultry accordingly—as you would for cooking fresh. Choose younger birds for roasting, frying, and broiling—more mature birds for braising and stewing. Meat from older, less tender birds can go into dishes that use cooked meat.

PACKAGING MATERIALS

Wrapping materials, bags, and rigid containers—all made especially for freezer storage—are the materials used for packaging.

Wrapping materials.—Moisture-vapor-resistant materials suitable for wrapping uncooked poultry and cooked poultry meat include: Aluminum freezer foil (heavy weight); cellophane-coated freezer paper; laminated freezer paper (two or more layers of materials held together by a binding compound); polyethylene-coated freezer paper; polyethylene plastic; and, for short storage periods only, wax-coated freezer paper.

Most of these wrapping materials are available in both rolls and sheets.

Aluminum foil is sealed by holding edges together and folding them over two or three times. Freezer paper is best sealed with special freezer tape, which sticks well at low temperatures; it may also be tied with string, or secured with a good-quality rubber band. Some of the wrapping materials may be heat-sealed with a regular household iron or a special sealing iron.

Ordinary waxed papers are not sufficiently moisture-vapor-resistant for packaging foods to be frozen.

Bags.—Bags of different kinds of plastic or of coated or laminated freezer paper for packaging whole or half birds are available in various sizes. Other freezer bags of moisture-vapor-resistant materials are suitable for packaging cut-up uncooked poultry, cooked poultry meat, and cooked poultry dishes that contain little or no liquid. They may be used for cooked dishes that are semiliquid, but they are not as convenient for these foods as rigid containers.

Some freezer bags may be heat-sealed. Some can be sealed by twisting the top, then folding it back and securing it with a string or a rubber band that will withstand freezer temperatures. Some come with paper-covered metal strips for sealing.

Rigid containers.—Rigid containers are best for cooked-poultry dishes that are liquid or semiliquid. They may be of such moisture-vapor-proof materials as aluminum, glass, glazed pottery, plastic, or plain or enameled tin, or of moisture-vapor-resistant materials, such as heavily waxed cardboard.

Covers of these containers should fit tightly. If they do not, it's a good idea to place freezer tape where container and cover join. Unwaxed or lightly waxed cartons used for cottage cheese and ice cream are not sufficiently moisture-vapor-resistant to use for freezing.

Some rigid containers can also be used for reheating and serving the food. Most rigid containers have a mark or line to show how high food should come in the package—to allow for the expansion of liquid during freezing.

Packaging accessories.—Among the items that you may find useful when packaging are: A special sealing iron for heat-sealing bags or wraps; paper or special heat-resistant material to use between the iron and plastic bags; special marking pencils for labeling packages; stockinette tubing, which is used as an outer covering to protect freezer wrappings; a special stand and funnel for filling certain types of freezer containers with semiliquid foods; and a small wooden block or box to use when heat-sealing packages.

Care of packaging materials.—Protect packaging materials from dust and insects. Keep bags and rolls of wrapping materials that may become brittle in a place that is cool and not too dry. Use them before they dry out from long storage.

GENERAL FREEZING PROCEDURE

Preparation for freezing

Keep the product and anything that touches it clean. Be particularly careful with cooked poultry as with all cooked meats. Bacteria, which cause spoilage, are spread by handling. Mixtures that contain sauces and gravies are especially suitable for the growth of bacteria.

Prepare poultry for packing and pack according to specific directions given for the product you are freezing, pages 7 to 23.

When packaging poultry in wrappings or bags, press the air out of the package before sealing.

When packaging poultry or poultry dishes in rigid containers, pack tightly to force as much air out of the food as possible. Then, for liquid and semiliquid foods, allow the amount of headspace recommended in the guide at upper right.

Keep sealing edges free from food and moisture. Seal carefully according to manufacturer's directions. When heat-sealing, regulate heat of iron carefully; too much heat melts or crinkles the material and prevents sealing.

Label each package with the date, kind of bird, type of product, and weight or number of servings or pieces.

Freezing

Freeze poultry at 0° F. or below. Freeze it quickly to maintain highest quality in the food. To speed freezing, place packages against the sidewalls or on metal freezing plates or shelves and leave a little space between packages so air can circulate freely.

If you try to freeze too many packages at once, the rate of freezing will be slowed to the point that food may spoil or lose quality before it is frozen.

Headspace To Allow

	Pint	Quart
Bag............	½ inch.	1 inch.
Low, broad container.	¼ inch.	½ inch.
Tall, straight or flared container.	½ inch.	1 inch.

To avoid spoilage and loss of quality, put no more unfrozen food in a freezer than will freeze solid within 24 hours. Usually this will be about 2 or 3 pounds of food to a cubic foot of freezer capacity.

Storing

Store frozen poultry at 0° F. or below. Keep freezer temperature uniform; constantly changing storage temperatures may dry out the food and hasten the development of off-flavors.

The maximum length of time a frozen poultry product should be stored depends on the quality of the poultry at the time of freezing, packaging material used, the amount of air in the package, and storage temperature. Maximum storage times for home-frozen poultry that has been wrapped, frozen, and stored under the most favorable conditions are:

Uncooked poultry:	Months
Chicken and turkey.........	12
Duck and goose...........	6
Giblets.................	3
Cooked poultry:	
Slices or pieces—	
Covered with broth or gravy................	6
Not covered with broth or gravy.............	1
Sandwiches of poultry meat.	1
Cooked-poultry dishes......	6
Fried chicken.............	4

It's a good idea to post a list of frozen foods near the freezer and to keep it up to date. List each food as you put it in—the form in which it was frozen, date of freezing, and the number of packages; check each package off as you take it out. Check this list from time to time so that you will use frozen foods before they begin to lose quality.

In case of emergency

If power fails or the freezer stops operating normally for any other reason, try to determine how long before the freezer will be back in normal operation.

A fully loaded freezer usually will stay cold enough to keep foods frozen for 2 days if the cabinet is not opened. In a cabinet with less than half a load, food may not stay frozen more than 1 day.

If normal operation cannot be resumed before the food will start to thaw, use dry ice. If dry ice is placed in the freezer soon after the power is off, 25 pounds should keep the temperature below freezing for 2 to 3 days in a 10-cubic-foot cabinet with half a load, 3 to 4 days in a loaded cabinet.

Place the dry ice on cardboard or small boards on top of packages and do not open freezer again except to put in more dry ice or to remove it when normal operation is resumed.

Do not handle dry ice with bare hands; it can cause burns. Be sure room is well ventilated before you use it.

Or move food to a locker plant, using insulated boxes or thick layers of paper to keep it from thawing.

Refreezing

Frozen raw or cooked poultry that has thawed may be safely refrozen if it still contains ice crystals or if it is still cold—about 40° F.—and has been held no longer than 1 or 2 days at refrigerator temperatures after thawing. Thawing and refreezing may lower the eating quality of the food.

FREEZING UNCOOKED POULTRY

Cleaning the bird

Ready-to-cook birds should need little cleaning. Inspect for pinfeathers and for any lung tissue that may have been left inside of the bird. Wash the bird inside and out.

To clean dressed birds, first remove pinfeathers. Then singe off hairs and cut off head and feet. Cut out oil sac, which is on top of the tail.

Next, cut circle around vent below tail, freeing it for removal with internal organs. Make a crosswise slit large enough for drawing between this circle and rear end of breastbone. If the bird is to be frozen whole, leave a band of skin between the two cuts.

Remove internal organs and the vent.

Slit skin lengthwise at the back of the neck. Then slip this skin down and remove crop and windpipe. Cut neck off short and save it; leave neck skin on the bird. Wash bird in cold running water.

Trim and cut blood vessels from heart. Cut away the green gall sac from the liver, being careful not to break the sac; gall will spoil the flavor of any meat it touches. Cut through one side of gizzard to inner lining. Remove and discard lining and contents. Wash giblets.

After cleaning fresh-killed poultry, leave it in the refrigerator below 40° F. for 12 hours before freezing—to insure maximum tenderness.

Home Freezing of Poultry (5) 173

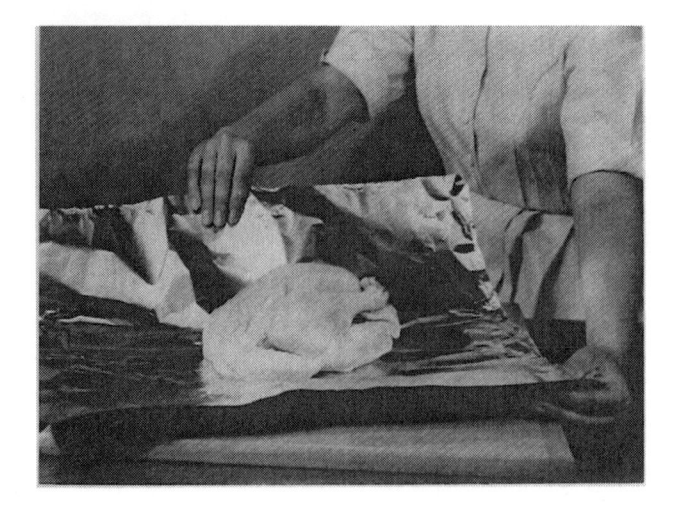

Place bird (legs tied together, wings pressed close to the body) in center of wrapping sheet—long way of the bird, long way of sheet.

WRAPPING A WHOLE BIRD—"LOCKER" WRAP

Whole birds

Chickens, ducks, geese, and turkeys may all be frozen whole, ready to roast.

Do not stuff birds before freezing. The time it takes to cool stuffing in the bird during freezing and to thaw and reheat it may be long enough to permit growth of food-spoilage and food-poisoning bacteria. Commercially stuffed frozen poultry is prepared under special conditions that cannot be duplicated in the home.

To freeze.—Tie leg ends of cleaned bird together. Press the wings close to the body.

It is best to wrap and freeze giblets separately, but they may be frozen with the bird if it is to be used within 3 months. If they are to be frozen with the bird, wrap them in moisture-vapor-resistant material and place them in the body cavity.

Place bird in freezer bag or wrap it in moisture-vapor-resistant material as shown at right and above. Various kinds of freezer papers can be used in the same way as the foil shown here.

Seal and freeze.

If giblets are to be frozen separately, place them in pliofilm bags or wrap them in moisture-vapor-resistant material. Seal and freeze.

To thaw.—Keep bird frozen until time for cooking. It is best to thaw the bird in the refrigerator (see timetable, p. 9); thaw bird on a tray until bird is pliable. To speed thawing, remove wrapper before putting bird in the refrigerator or immerse bird in a watertight bag in cold water. Change water often. If desired, the bird may be partially thawed in the refrigerator, then finished in cold water.

(*Continued on page 9*)

Bring long sides of sheet over bird and fold together about 1 inch of the edges. Fold again as necessary to bring sheet tight and flat on top of the bird.

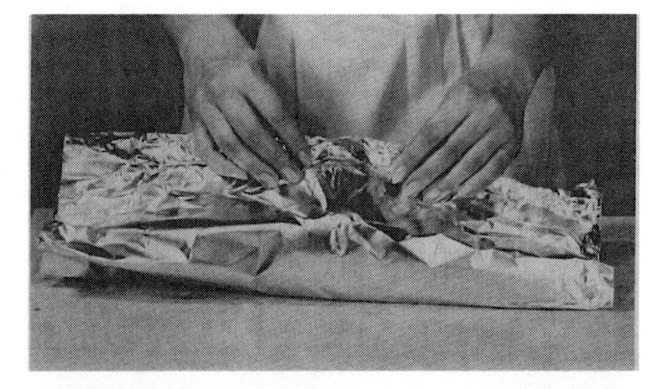

Press wrapping close to the bird to force air out.

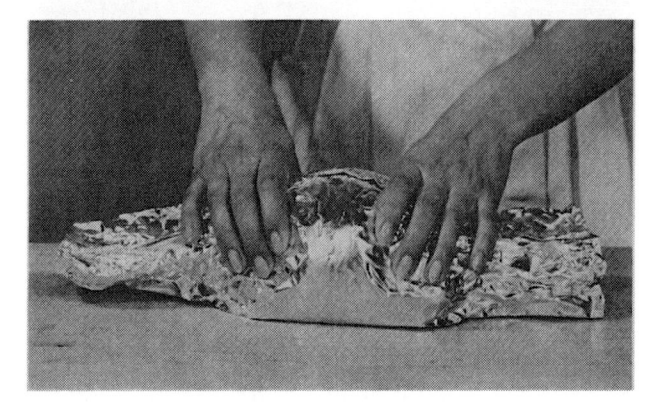

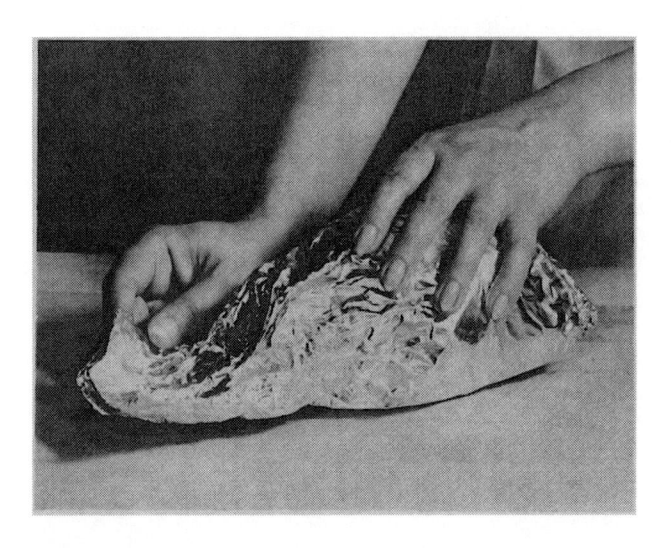

At each end, fold corners toward each other. Fold ends upward and over until package is tight.

Home Freezing of Poultry (7) 175

CHICKEN HALVES

Remove neck-and-back strip by cutting from neck to tail along both sides of the backbone.

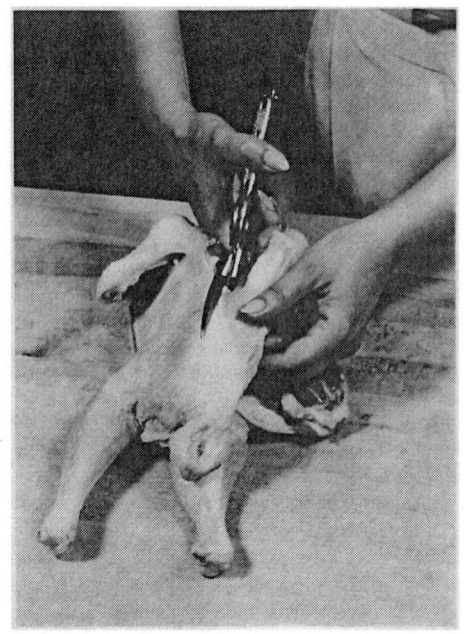

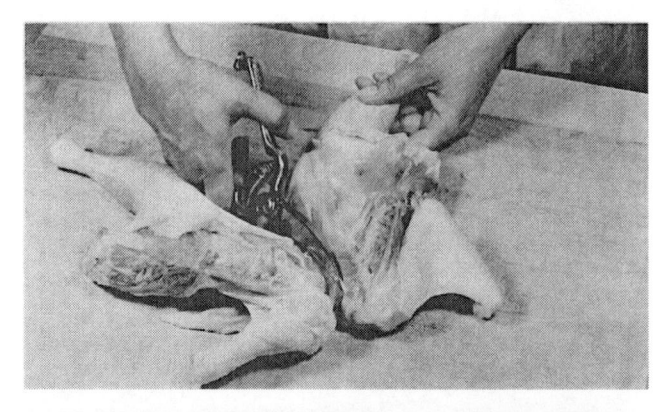

Turn bird on its breast. Cut in two along the breastbone. Or, if desired, breastbone may be removed as shown on page 12 before breast is cut in two.

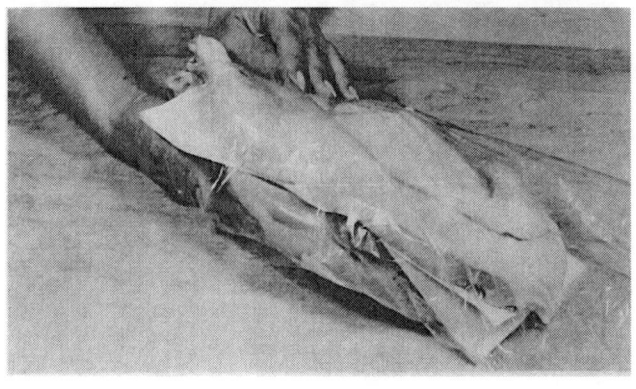

If halves are packaged together, place a double fold of freezer paper between them.

Whole birds—Continued

Timetable for Thawing Whole Birds in the Refrigerator

Chickens:	
4 pounds and over.	1 to 1½ days.
Less than 4 pounds.	12 to 16 hours.
Ducks, 3 to 5 pounds.	1 to 1½ days.
Geese, 4 to 14 pounds.	1 to 2 days.
Turkeys:	
18 pounds and over.	2 to 3 days.
Less than 18 pounds.	1 to 2 days.

Prepare thawed birds for roasting in the same way as unfrozen poultry.

Half birds—chickens or turkeys

Lay the bird on its side and remove neck-and-back strip by cutting from neck to tail along both sides of the backbone. Then lay bird on its breast, open it, and cut along the inside of the breastbone.

To freeze.—Package halves together as a whole bird, or package each half separately. If halves are packed together, place a double piece of freezer paper between them.

It is best to wrap and freeze giblets separately, but they may be frozen with the halves if the halves are to be used within 3 months. If giblets are to be frozen with the halves, wrap them in moisture-vapor-resistant material and place them in the body cavity.

Place half or halves in freezer bag or wrap in moisture-vapor-resistant material. Seal and freeze.

If giblets are to be frozen separately, place them in pliofilm bags or wrap them in moisture-vapor-resistant material. Seal and freeze.

To thaw.—Follow directions for whole birds. It will take 1 to 2 days to thaw large turkey halves, 3 to 9 hours to thaw chicken halves in the refrigerator.

Cut-up chicken or turkey

Chicken or turkey to be fried, fricasseed, stewed, or braised takes up less space in the freezer and is in more convenient form for use if cut up before freezing.

Disjoint cleaned chicken as shown on pages 10 to 12. If preferred, the leg may be left whole.

Disjoint turkeys in the same way, except wing may be left on each lengthwise half of the breast if preferred.

After bird has been cut up, wash the pieces in cold water. Dry.

Separate meaty pieces (breast, legs, wings) from bony pieces (neck, back, wingtips). Use bony pieces for broth (p. 14).

To freeze.—Place meaty pieces close together in a freezer bag or carton. Or wrap in moisture-vapor-resistant material as shown for whole birds on pages 6 and 7. To make thawing quicker, place each piece of poultry in a fold of freezer wrapping material. Seal and freeze.

It is best to wrap and freeze giblets separately. Freeze with meaty pieces only if they will be used within 3 months.

If giblets are to be frozen separately, place them in pliofilm bags or wrap them in moisture-vapor-resistant material. Seal and freeze.

To thaw.—Unwrap package. Separate pieces so air can reach each piece. Leave in refrigerator during thawing. Complete thawing will take 3 to 9 hours.

Chicken for stewing or braising may be cooked without thawing.

DISJOINTING A CHICKEN

One method of disjointing a chicken is shown here. Turkeys may be disjointed in the same way, except the wing may be left on each half of the breast if preferred.

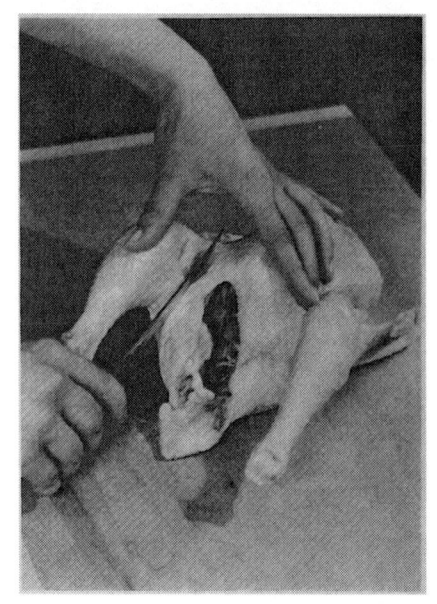

1. Cut skin between thighs and body of the bird.

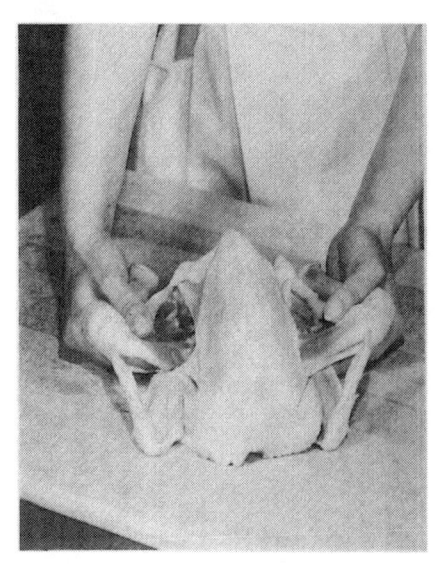

2. Grasp a leg of the bird in each hand and lift the bird from the table, bending its legs back as you lift. Bend legs until hip joints are free.

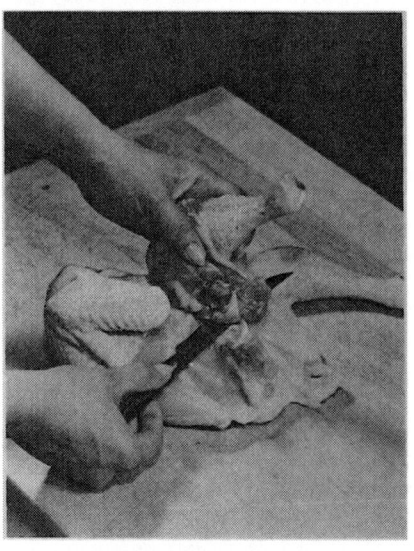

3. Remove leg-and-thigh piece from one side of the body by cutting from back to front as close as possible to the bones in the back of the bird.

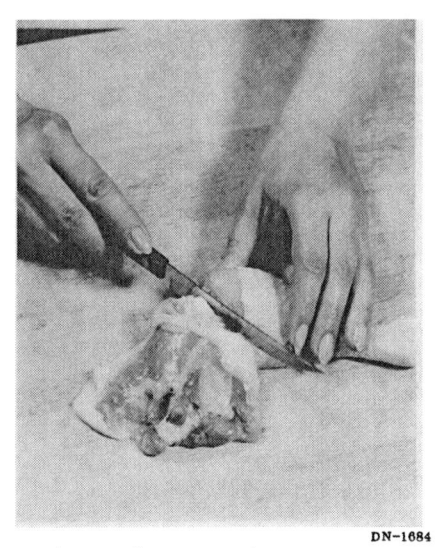

4. Locate knee joint by squeezing thigh and leg together. Cut through this joint to separate thigh and leg.

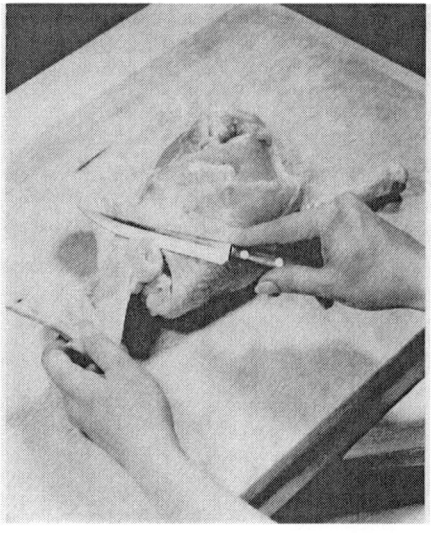

5. Remove wing from body. Start cutting on inside of wing just over the joint. Cut down and around the joint.
To make wing lie flat, either cut off the wingtip or make a cut on the inside of the wing at the large wing joint; cut just deep enough to expose the bones.

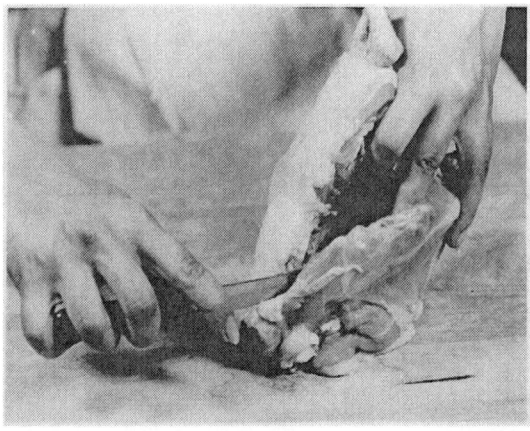

6. Turn body of bird over and remove other leg-and-thigh piece and other wing in the same way. Separate thigh and leg at the knee joint.

7. Divide the body by placing bird on neck end and cutting through the meat from back to tail along the end of the ribs. Then cut along each side of the backbone through the rib joints, then between backbone and flat bone at the side of the back. Cut the skin that attaches neck-and-back strip to the breast.

(See next page.)

Home Freezing of Poultry (11) 179

8. Place neck-and-back strip, skin side up, on cutting board. Cut strip in two just above the spoon-shaped bones in the back.

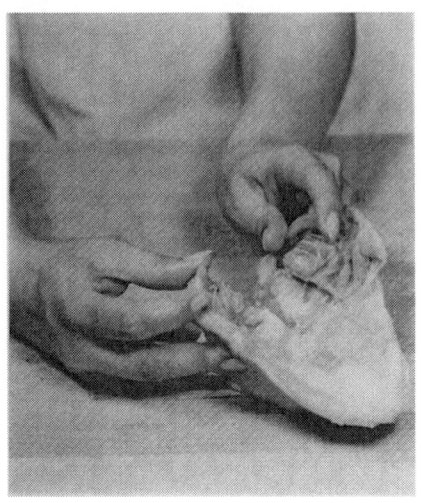

9. Place the breast, skin side down, on the cutting board. Cut through the white cartilage at the V of the neck as shown.

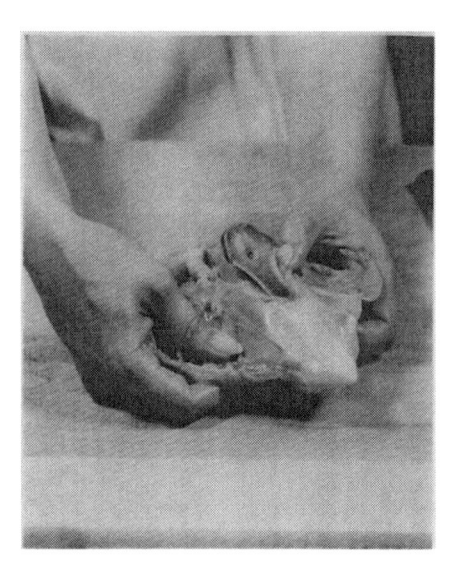

10. Grasp the breast piece firmly in both hands. Bend each side of the breast back and push up with the fingers to snap out the breastbone. Cut the breast in half lengthwise.

11. The disjointed chicken. Meaty pieces at left—legs, thighs, wings, breast halves. Bony pieces at right—wingtips, back strip, back, and neck.

FREEZING COOKED POULTRY

Cooked poultry for freezing can come from birds cooked especially for that purpose—or from leftovers.

Any kind of cooked poultry can be frozen satisfactorily. But some—fried chicken, slices or pieces packed without broth or gravy, and sandwiches—have a relatively short storage life (p. 4).

Meat from cooked whole birds should be removed from the carcass and packed solidly to eliminate air, which causes development of rancid off-flavor. Addition of broth or gravy will prevent contact with air and lengthen the storage life.

Roast poultry

If you roast the bird specially for freezing, do not stuff it. Make the stuffing later when the bird is cooked for eating.

Cool roast bird quickly in the refrigerator. When it is cool, remove meat from the bones. Remove as much of it as possible in large pieces or in slices for special use.

To freeze.—Divide meat into large pieces or slices, and small pieces. Package separately in meal-sized amounts.

Package large pieces in freezer bags, or wrap in moisture-vapor-resistant wrapping paper. Seal and freeze.

Package slices or small pieces covered with gravy or broth in rigid containers, leaving headspace (p. 4). Package them without gravy or broth in freezer containers or bags or moisture-vapor-resistant wrapping material. Seal and freeze.

To thaw.—Thaw meat frozen without gravy or broth in the refrigerator in the package.

Thaw meat frozen with broth or gravy in the package until it can be removed easily. For this partial thawing, either put package in the refrigerator or dip it in hot water.

Remove from the package and heat in the oven at 400° F. (hot oven) or on top of the range. If you heat on top of range, start with low heat.

Packaging large slices of roast turkey.

Home Freezing of Poultry (13) 181

Adding broth to small pieces of turkey packed for freezing.

Stewed, braised, or fricasseed poultry

If you stew, braise, or fricassee poultry especially for freezing, you may prefer to save the uncooked bony pieces to be used for broth (upper right).

Cool cooked poultry quickly in the refrigerator.

To freeze.—Package bony and meaty pieces together covered with broth or gravy in rigid containers. Leave headspace (p. 4). Seal and freeze.

Or remove meat from the bones and package it with broth in rigid containers or without broth in cartons, freezer bags, or moisture-vapor-resistant wrapping material. Leave headspace if broth is added. Seal and freeze.

Or, if you prefer, separate bony and meaty pieces and pack them in one of the ways given above. Seal and freeze.

To thaw.—Thaw as for roast poultry (p. 13).

Chicken or turkey broth

Simmer bony pieces of chicken or turkey in water to cover until meat is loosened from the bones.

Remove bones and skin. Skim off excess fat.

To freeze.—Remove pieces of meat or leave with broth as desired. Cool quickly. Pack broth in freezer containers, leaving headspace (p. 4). Seal and freeze.

To thaw.—Heat frozen broth in saucepan on top of the range.

Sandwiches

Most kinds of poultry sandwiches freeze satisfactorily.

Make the filling of sliced or chopped chicken or turkey meat. Mix chopped meat with cream cheese or a small amount of mayonnaise or salad dressing to moisten the ingredients. Add chopped olives or pickles, grated onion, or diced celery if desired. Do not add such raw vegetables as lettuce, tomatoes, or watercress. These lose crispness, color, and flavor when frozen.

Use day-old bread for sandwiches; it is better than fresh bread for sandwiches to be frozen. Spread bread with butter, margarine, or cream cheese. Spread liberally before adding chopped-meat filling—to keep the filling from soaking through on the bread.

To freeze.—Wrap each sandwich in a double thickness of heavy waxed paper (for storage no longer than 1 week) or in moisture-vapor-resistant wrapping material (for storage up to 1 month). Seal and and freeze.

If you wish, place wrapped sandwiches in a cardboard box to keep them from being crushed during freezer storage.

To thaw.—Thaw sandwiches in the original wrapping at room temperature. It will take 3 to 4 hours. Use soon after thawing.

FREEZING COOKED-POULTRY DISHES

Creamed dishes, casseroles, salads, pies, and various other main dishes made with cut-up cooked chicken or turkey are suitable for freezing.

Preparing the dish

The procedure that will make the best frozen product differs somewhat for the different types of food, as shown in the recipes given on pages 17 to 23.

In general, prepare the food as you would if it were to be served immediately. Do not overcook.

Some ingredients are better if slightly undercooked before freezing. Vegetables and macaroni, for example, may be too soft and have a warmed-over taste when reheated if cooked well done before freezing.

Pastry crusts frozen unbaked are more tender and flaky and have a fresher flavor than those baked before freezing.

Crumb or cheese toppings are better if added when food is reheated for serving than if they are frozen with the dish.

For thickening sauces and gravies, use waxy rice flour (also called sweet rice flour or mochiko) if it is available. A sauce thickened with this flour is not so likely to separate and curdle or become watery when it thaws as one made with wheat flour.

Cooling

Cool food immediately after it is cooked. Because cooling stops the cooking it helps keep the natural flavor, color, and texture of the food. It also retards or prevents the growth of bacteria that may cause spoilage.

To cool, set uncovered pan of food in iced or very cold water; change the water to keep it cold. Or set pan on ice or put it in a cold place.

Packaging and freezing

Package cooled food *promptly* in freezer containers or moisture-vapor-resistant packaging material. Most cooked-poultry dishes, because they are liquid or semiliquid, should be packaged in rigid containers. Allow headspace for these foods as recommended on page 4.

Foods that are to be baked after freezing can be used more conveniently if they are frozen in ovenproof containers. Or, a food may be frozen in the casserole in which it will be baked, then removed after freezing and wrapped for freezer storage. Line the casserole with heavy aluminum foil before freezing so the food can be removed easily. Remove the foil and replace food in casserole for reheating.

Package other liquid and semiliquid foods in freezer containers that have a wide top opening, so the food can be removed for reheating before it is completely thawed.

Choose a container of a size that holds only enough for one meal for your family. Quart containers hold 4 to 6 servings; pints, 2 to 3. If you use quart containers you may want to separate the food into two or three layers with a double thickness of water-resistant material such as cellophane; the smaller blocks of frozen food can be reheated in a shorter time than one large block.

Reheating

Food in ovenproof containers.—Take container from freezer and put directly into a hot oven to reheat.

Home Freezing of Poultry (15) 183

Food in other containers.—If it is necessary to partially thaw food before removing it from other rigid containers—in order to remove cellophane between layers or to remove food from the container—place the package in lukewarm water for a few minutes. Or allow the food to stand in the container at room temperature for 30 to 60 minutes. It may also be necessary to thaw food slightly if you are going to cook it over direct heat.

Reheat soups and other dishes that contain a high percentage of liquid in a saucepan over direct heat. Keep heat low. Stir gently to prevent sticking.

Reheat other dishes in a double boiler or in a hot oven. Reheating in the oven is slower, but takes little or no attention.

If you reheat food in a double boiler, start with warm water in the bottom pan. Break up the frozen mass as it thaws. Stir gently only when necessary to keep food from sticking; too much stirring may result in less desirable texture. Large blocks of frozen food or several small blocks will heat faster in a wide-bottom saucepan set in another pan of hot water.

If you prefer to thaw food completely before reheating, thaw in the refrigerator. Thawing at room-temperature may cause dangerous spoilage if it takes more than 3 or 4 hours. *Once the food is thawed, reheat it and use it immediately.*

Cool cooked food immediately by placing pan on ice or in iced water. Then package and freeze cooled food promptly.

Recipes

The following recipes include directions for both freezing and reheating. Recipes yield from 16 to 25 portions.

Although each recipe specifies cooked chicken or turkey, either may be used in any of the recipes. For added flavor, fresh chicken fat may be used to replace butter or margarine.

The table below will help you in figuring the approximate amount of cooked meat (without skin or fat) you'll get from birds of different sizes.

Weight of bird, ready to cook	Amount of chopped cooked meat
Chicken:	Quarts
3 lb.	1
4 lb.	1¼
Turkey:	
5 lb.	1¾
8 lb.	2¾
12 lb.	4½
18 lb.	6¾

Chicken-corn casserole

25 portions, ⅔ cup each

½ cup butter or margarine
1¼ cups sifted flour
1½ quarts chicken broth
1½ teaspoons salt
½ teaspoon pepper
2 tablespoons finely chopped onion
1¾ quarts chopped cooked chicken
1½ quarts drained cooked or canned whole-kernel corn
1½ cups grated American cheese
4-ounce can pimiento, chopped

Melt the fat, blend in the flour, and stir in the broth. Add salt, pepper, and onion. Cook until thickened, stirring constantly. Combine sauce, chicken, corn, cheese, and pimiento.

To freeze.—Cool food quickly. Pack in freezer containers, leaving headspace (p. 4). Seal and freeze immediately.

To prepare for serving.—If food is frozen in an ovenproof container, uncover and top with dry breadcrumbs mixed with butter or margarine. If the food is transferred to a baking dish for reheating, add the crumb topping after 30 minutes' baking or when food has thawed enough to press into shape of the baking dish. Bake at 400° F. (hot oven) about 1 hour for pints, 1 hour 45 minutes for quarts—or until food is heated through and crumbs are golden brown.

Turkey chop suey

24 portions, ⅔ cup each

1¼ quarts celery cut in 1-inch strips
1½ cups sliced onion
1 tablespoon salt
¼ teaspoon pepper
2 tablespoons sugar
2 quarts turkey broth and bean sprout liquid
1½ cups canned bean sprouts
1½ cups sliced water chestnuts, if desired
2 quarts cooked turkey in pieces or 2-inch strips
¼ cup fat or oil
¾ cup cornstarch
½ cup cold water
½ cup soy sauce

Simmer the celery, onion, salt, pepper, and sugar in the broth and liquid from the bean sprouts for 20 minutes. Heat the turkey in the fat and add it, bean sprouts, and water chestnuts to the vegetable mixture. Blend cornstarch with cold water and stir into the mixture. Simmer for 15 minutes, stirring frequently. Add soy sauce.

To freeze.—Cool the food quickly. Pack in freezer containers, leaving headspace (p. 4). Seal and freeze immediately.

To prepare for serving.—Thaw and reheat in double boiler—30 minutes for pints, 1 hour for quarts.

Turkey pie

24 portions, ¾ cup each

1 cup butter or margarine
1¾ cups sifted flour
1 tablespoon salt
1½ quarts hot turkey broth
1½ quarts hot milk
3 quarts diced cooked turkey

Pastry Topping

3 cups sifted flour
2 teaspoons baking powder
1 teaspoon salt
1 cup shortening
½ cup cold water

Melt fat and blend in the flour and salt. Add broth and milk. Cook until thickened, stirring constantly. Add the turkey.

For the topping, sift together the flour, baking powder, and salt. Cut in the shortening until the mixture is granular. Add water and mix lightly with a fork. Turn the dough out onto a lightly floured board or pastry cloth, roll to about one-eighth inch in thickness, and cut into serving-size pieces.

To freeze.—Cool the creamed turkey quickly. Pack in ovenproof freezer containers and place pastry pieces on top. Leave headspace (p. 4). Seal and freeze immediately.

Or freeze pastry separately. Place two layers of waxed paper between layers of pastry pieces so the frozen pieces can be separated easily, then wrap in freezer paper.

To prepare for serving.—Bake at 400° F. (hot oven) for 1 hour if packed in pint containers. If pastry was frozen separately, bake creamed turkey for 40 minutes at 400° (hot oven). Then turn the oven to 425° and place the pieces of pastry on top; bake for another 25 minutes.

Adding pastry topping to turkey pie.

Turkey goulash

24 portions, ¾ cup each

1 pound shell macaroni
2 quarts boiling salted water
1½ quarts hot turkey broth
2 cups cooked tomatoes
¼ cup tomato paste
½ cup chopped onion
1½ teaspoons chopped green pepper
¾ cup fat or oil
1½ cups sifted flour
1 teaspoon salt
2 quarts chopped cooked turkey

Cook macaroni in boiling salted water for 15 minutes, or until almost tender. Drain and rinse. Mix broth, tomatoes, tomato paste, onion, and green pepper. Heat fat and blend in flour and salt; add to broth mixture. Cook until thickened, stirring frequently. Add turkey and cooked macaroni.

To freeze.—Cool quickly. Pack in ovenproof freezer containers, leaving headspace (p. 4). Seal and freeze immediately.

To prepare for serving.—Bake at 400° F. (hot oven)—pints for 1 hour, quarts for 1¼ hours. A topping of crumbs mixed with butter or margarine may be added one-half hour before baking is finished.

Turkey pimiento soup

24 portions, 1 cup each

1 quart chopped celery
¼ cup chopped onion
¾ cup butter or margarine
2 cups sifted flour
2 tablespoons salt
¾ teaspoon pepper
3¼ quarts hot turkey broth
3 cups chopped cooked turkey
2 quarts hot milk
⅓ cup chopped pimiento

Cook celery and onion in the fat until light brown (about 15 minutes). Blend in the flour, salt, and pepper. Gradually stir in the turkey broth. Cook over low heat until slightly thickened, stirring occasionally. Add the turkey and continue cooking for 15 minutes. Add milk and pimiento to the mixture. Reheat.

This soup may also be frozen without the milk and pimiento.

To freeze.—Cool the food quickly. Pack in freezer containers, leaving headspace (p. 4). Seal and freeze immediately.

To prepare for serving.—Heat in double boiler.

If soup is frozen with milk and pimiento, allow about 30 minutes for pints to thaw and reheat, 50 minutes for quarts.

If soup is frozen without milk and pimiento, allow about 45 minutes for quarts to thaw, add these ingredients, and allow 20 minutes more to heat for serving.

Curried turkey and ham

24 portions, ½ cup each

⅓ cup chopped green pepper
3 tablespoons chopped onion
½ cup fat or oil
1 cup sifted flour
2½ teaspoons salt
⅛ teaspoon pepper
1½ teaspoons curry powder
2¼ quarts hot milk
1 quart diced cooked turkey
2 cups diced cooked cured ham

Cook green pepper and onion lightly in the fat or oil. Stir in the flour and seasonings. Gradually blend into milk, stirring constantly. Cook until thickened. Add turkey and ham to sauce.

To freeze.—Cool the food quickly. Pack in freezer containers, leaving headspace (p. 4). Seal and freeze immediately.

To prepare for serving.—Remove from freezer containers and reheat in a double boiler 45 minutes for pints, 1 hour for quarts. Or reheat in a saucepan over low heat. Serve on rice, Chinese noodles, biscuits, or toast.

Barbecued turkey

24 portions, ½ cup each

2 tablespoons fat or oil
¼ cup vinegar
1⅓ cups water
2⅔ cups chili sauce
½ cup chopped onion
2 cups chopped celery
1 tablespoon powdered dry mustard
1 tablespoon salt
¼ cup brown sugar
2½ quarts chopped cooked turkey

Combine fat, vinegar, water, chili sauce, onion, and celery. Add the mustard and salt mixed with the sugar. Heat, but do not cook enough to soften the vegetables. Add turkey to sauce. Reheat.

To freeze.—Cool the food quickly. Pack in freezer containers, leaving headspace (p. 4). Seal and freeze immediately.

To prepare for serving.—Reheat in double boiler—30 minutes for pints and 1 hour for quarts. Serve between halves of buttered rolls.

Potato surprise

24 portions, ½ potato each

12 large baked potatoes
Butter or margarine
Salt
Pepper
Hot milk (about 1½ cups)
¾ cup chopped onion
1½ cups grated cheese
½ cup fat
½ cup sifted flour
1 teaspoon salt
2¼ cups turkey broth
1½ quarts chopped turkey

Cut large baked potatoes in half lengthwise as soon as they are taken from the oven. Scoop out inner portion and mash. Season with butter or margarine, salt, and pepper. Add enough milk to moisten. Beat the mixture until it is smooth and fluffy and add the onion.

Stuff the shells with two-thirds of the potato mixture. Mix the remainder of the potato mixture with the cheese.

Melt the fat and blend in the flour and salt. Add turkey broth gradually, stirring constantly until thickened. Add the turkey.

Make an indentation in the potato filling and fill it with a spoonful of the turkey mixture. Cover the potato halves with the potato-cheese mixture.

To freeze.—Cool the food quickly. Wrap and seal for freezing and freeze at once.

To prepare for serving.—Unwrap, and bake at 425° F. (hot oven) for 55 minutes or until filling is hot and the top is browned.

Turkey with pineapple

24 servings, ½ cup each

½ cup fat or oil
½ cup sifted flour
2 cups hot milk
2 cups hot turkey broth
1 quart cooked turkey, cut in 1-inch cubes
2 cups cooked cured ham, cut in 1-inch cubes
2 cups canned pineapple pieces
2 tablespoons of pimiento strips
Salt

Melt the fat and blend in the flour. Stir into the combined milk and broth and cook until thickened, stirring constantly. Add turkey, ham, pineapple, and pimiento. Season to taste. Heat to blend flavors.

To freeze.—Cool quickly. Pack in freezer containers, leaving headspace (p. 4). Seal and freeze at once.

To prepare for serving.—Remove from containers and reheat in a double boiler—30 minutes for pints, 1 hour for quarts.

Turkey turnovers

24 turnovers

Filling

1½ quarts finely chopped cooked turkey
½ cup finely chopped onion
2 cups canned concentrated cream of chicken or mushroom soup
¼ cup chopped parsley
1 teaspoon salt
½ teaspoon pepper
½ teaspoon poultry seasoning

Pastry

6 cups sifted flour
2 teaspoons salt
2 cups shortening
Cold water, about 1 cup

Combine all ingredients for the filling and mix well.

For the pastry, sift together the flour and salt. Cut in the shortening until the mixture is granular. Add just enough water to moisten; mix quickly. Roll pastry to ⅛-inch thickness on a lightly floured board. Cut into circles 5 inches in diameter.

Place 2 to 3 tablespoons of filling on one-half of each pastry circle. Moisten edge and fold the other half of the pastry over the filling; seal the edges with a fork.

To freeze.—Wrap and seal for freezing and freeze immediately.

To prepare for serving.—Place turnovers on a cooky sheet (unless they were frozen in ovenproof containers) and bake at 425° F. (hot oven) for 40 minutes, or until pastry is browned. Serve with peas in a cream sauce if desired.

Packaging turnovers for freezing.

Creamed chicken

16 portions, ¾ cup each

¾ cup butter or margarine
1⅛ cups sifted flour
1 tablespoon salt
1 quart hot chicken broth
1 quart hot milk
2 quarts diced cooked chicken

Melt fat and blend in flour and salt. Add broth and milk. Cook until thickened, stirring constantly. Add the chicken and heat to blend flavors.

To freeze.—Cool food quickly. Pack in freezer containers, leaving headspace (p. 4). Seal and freeze immediately.

To prepare for serving.—Heat frozen creamed chicken in top of a double boiler. For pints, it takes about 30 minutes.

Scalloped turkey with rice

24 portions, ⅔ cup each

2 cups rice (uncooked)
1 quart salted water
1½ quarts cooked chopped turkey
2 cups mushrooms
¾ cup butter or margarine
¾ cup sifted flour
2 teaspoons salt
¼ teaspoon pepper
2 cups hot milk
1 quart hot turkey broth
¼ cup diced pimiento

Add rice to boiling salted water and cook over low heat for about 15 minutes. Remove from heat and let stand covered 5 to 10 minutes.

Arrange rice in alternate layers with turkey and mushrooms in casseroles or freezer containers, leaving headspace (p. 4).

Make a sauce by melting the fat and blending in the flour and seasonings and adding to the milk and broth. Cook until thickened, stirring constantly.

Pour just enough of the sauce into each casserole or freezer container to cover the ingredients. Sprinkle with pimiento. (Or, if desired, pimiento may be added just before baking.)

To freeze.—Cool the food quickly. Seal and freeze immediately.

To prepare for serving.—Bake at 400° F. (hot oven)—1 hour for pints, 1½ hours for quarts. Add pimiento before baking if it was not frozen. If the food is transferred to a baking dish, let it heat in the oven about 30 minutes, or until the food has thawed enough to press into shape of the baking dish, before adding pimiento.

Turkey harlequin

24 portions, ½ cup each

2 cups chopped onion
1 cup chopped green pepper
½ cup fat or oil
⅔ cup sifted flour
1 tablespoon salt
¼ teaspoon pepper
3 cups turkey broth
2 cups drained cooked tomatoes
2 cups apple juice
⅔ cup raisins
1 tablespoon chopped parsley
2 quarts diced cooked turkey

Cook onion and green pepper in fat or oil until tender but not brown. Stir in flour, salt, and pepper. Combine broth, tomatoes, apple juice, raisins, and parsley. Heat to boiling. Stir in the onion-green pepper mixture and cook over low heat until thickened, stirring frequently. Add turkey and reheat to blend flavors.

To freeze.—Cool quickly. Pack in freezer containers, leaving headspace (p. 4). Seal and freeze immediately.

To prepare for serving.—Remove from containers and reheat in double boiler, 30 minutes for pints and 70 minutes for quarts. Do not overstir. Serve over hot cornbread, biscuits, or rice.

Chicken a la king

16 portions, ¾ cup each

½ cup butter or margarine
6 tablespoons chopped green pepper
3 cups canned mushrooms
¾ cup sifted flour
2 teaspoons salt
3 cups hot chicken broth
3 cups hot milk
6 tablespoons finely cut pimiento
1½ quarts diced cooked chicken

Melt the fat and cook green pepper and mushrooms in it for about 5 minutes. Blend in the flour and salt, then add broth and milk. Cook until thickened, stirring constantly. Add pimiento and chicken. Heat to blend flavors.

To freeze. — Cool the food quickly. Pack in freezer containers, leaving headspace (p. 4). Seal and freeze immediately.

To prepare for serving.—Heat frozen chicken a la king in top of a double boiler. For pints, it takes about 30 minutes.

Molded turkey salad

24 portions, 1 by 4 by 4 inches

4 tablespoons (4 envelopes) unflavored gelatin
1 cup cold water
4 cans condensed cream of chicken soup
2 quarts finely diced cooked turkey
1 quart diced celery
1 cup chopped pimiento
1 pint cooked salad dressing

Soften gelatin in cold water for 5 minutes and add to hot condensed soup. Stir to dissolve gelatin; cool. Add turkey, celery, pimiento, and salad dressing. Pour into four 1-quart ring molds or into loaf pans or freezer containers. Chill until firm.

To freeze.—Dip ring mold in hot water to unmold; invert on rigid container and wrap in moisture-vapor-proof material. Or if in loaf pan or freezer container, cover tightly. Freeze immediately.

To prepare for serving.—Thaw at room temperature for 2 to 4 hours, depending on the size of the loaf. If frozen in a loaf pan or freezer container, after thawing dip in hot water to unmold.

Scalloped turkey with noodles

24 portions, ¾ cup each

1 pound (about 6 cups) dry noodles
3 quarts salted water
2 quarts cooked turkey meat, large pieces
¾ cup butter or margarine
¾ cup sifted flour
1 tablespoon salt
¾ teaspoon powdered dry mustard
1¼ quarts hot turkey broth
1¼ quarts grated or shredded American cheese

Cook noodles in boiling salted water for 9 minutes. Drain and rinse in cold water.

Arrange turkey and noodles in alternate layers in casseroles or ovenproof freezer containers.

Melt the fat and blend in the flour mixed with the salt and mustard. Stir into the broth. Cook until thickened, stirring constantly. Add cheese and stir until melted. Pour over turkey and noodles. Leave headspace (p. 4).

To freeze.—Cool the food quickly. Seal and freeze immediately.

To prepare for serving.—If the food is baked in an ovenproof freezer container, add crumbs mixed with butter or margarine and sprinkle with paprika before baking. If the food is transferred to a baking dish, let it heat in the oven until the food has thawed enough to press into the shape of the baking dish, before adding the topping. Bake at 400° F. (hot oven) about 1 hour for pints, 1½ hours for quarts or until food bubbles throughout.

INDEX TO FREEZING DIRECTIONS

Chicken— Page
  a la king 23
  and corn casserole 17
  braised 14
  broth 14
  creamed 22
  cut-up, ready to cook 9–12
  fricasseed 14
  halves, ready to cook 8, 9
  roast 13
  sandwiches 14
  stewed 14
  whole, ready to roast 7
Cooked poultry 13, 14
Cooked-poultry dishes 15–23
Duck—
  braised 14
  fricasseed 14
  roast 13
  stewed 14
  whole, ready to roast 7
General freezing procedure 4, 5
Goose—
  braised 14
  fricasseed 14
  roast 13
  stewed 14
  whole, ready to roast 7
Headspace to allow 4

  Page
Packaging materials 3
Selecting poultry to freeze 2
Storing 4
Refreezing 5
Turkey—
  barbecued 20
  braised 14
  broth 14
  chop suey 17
  curried, with ham 19
  cut-up, ready to cook 9–12
  fricasseed 14
  goulash 19
  halves, ready to cook 9
  harlequin 22
  pie 18
  potato surprise 20
  roast 13
  salad, molded 23
  sandwiches 14
  scalloped:
    with noodles 23
    with rice 22
  soup, turkey pimiento 19
  stewed 14
  turnovers 21
  whole, ready to roast 7
  with pineapple 20
Uncooked poultry 5–9

FREEZING MEAT AND FISH IN THE HOME

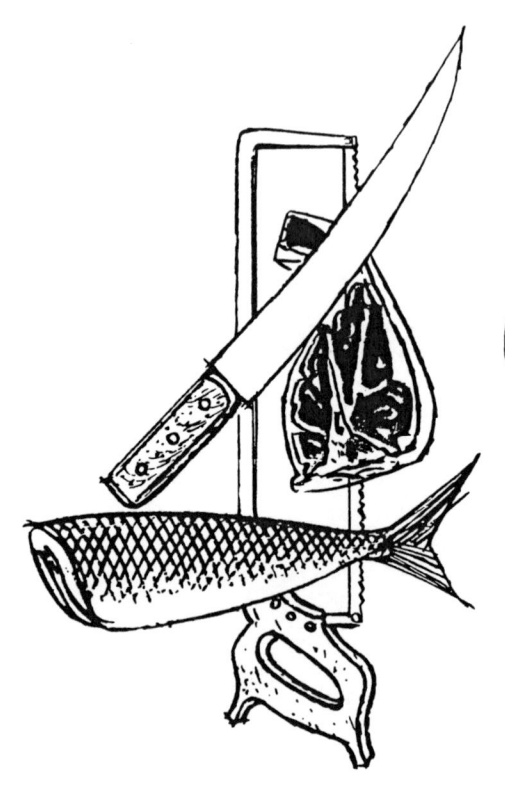

Contents

	Page
General freezing procedures	3
Cutting methods	4
Beef and veal	4
Pork	5
Lamb	5
Boning the cuts	6
Beef and veal	6
Pork	8
Lamb	10
Meat yields	12
Fish	14
Selection	14
Cleaning and dressing	15
Wrapping, freezing, and storing	18
Storage periods	20
Thawing meats and fish	21

FREEZING MEAT AND FISH IN THE HOME

Freezing is an excellent method of preserving meat and fish. It can be done simply and effectively at home if—

☐ Meat and fish are carefully selected, prepared, and packaged; and

☐ Home freezing equipment freezes quickly at 0° F. or lower, and maintains these temperatures for storage of frozen products.

Too high or constantly changing storage temperatures cause even properly packaged and frozen foods to lose quality and food value.

GENERAL FREEZING PROCEDURES

☐ Freeze only high-quality fresh meat and fish. Freezing does not improve quality.

☐ Keep all food to be frozen—and anything that touches it—clean. Freezing does not sterilize food; the extreme cold simply slows down changes that affect quality or cause spoilage in food.

☐ Thoroughly chill meat and fish (to below 40° F.) within 24 hours. Hang meat up to chill. Chill meat at a temperature of 33° to 36° F. Chill pork and veal, 1 to 2 days; beef and lamb, 5 to 7 days (to make more tender).

☐ Prepare and package in quantities to be cooked at one time.

☐ Protect all foods to be frozen by wrapping or packaging carefully in recommended materials. Proper packaging keeps foods from drying out and prevents air from entering the closed package and causing oxidation. Suit the packaging (wrapping materials, waxed cartons, bags, containers, etc.) to the kind, shape, and size of the food, and to its consistency—whether the pack is solid or liquid.

☐ Label each package with the date, kind and type of product, and weight or number of servings or pieces.

☐ Freeze quickly at 0° F. or lower. Turn the temperature control in the home freezer to the coldest position to speed up freezing and help prevent warming of stored frozen food. To freeze a maximum load and reduce it to storage temperature, wait 24 hours before returning the control to storage position; for a half load or less, 8 to 12 hours is sufficient.

☐ Limit the amount of food frozen at one time in a home freezer to 2 pounds per cubic foot of total storage space. Place the unfrozen packages against a refrigerated surface. Arrange them so that they do not touch any packages of frozen food stored in the freezer. If necessary, use a board or heavy cardboard, such as corrugated carton material, between the frozen and unfrozen packages to prevent them from coming in contact with each other.

☐ Store at 0° F. or lower.

☐ Limit storage time. See the table on page 20 for maximum storage times recommended for foods in moisture-vapor-proof or moisture-vapor-resistant wrappings.

☐ Cook thawed foods promptly.

CUTTING METHODS

BEEF AND VEAL

Cut the loin (4), rib (5), chuck (8), and round (2) into roasts or steaks. Cut the rump (3) into roasts. Cut the shanks (1), flank (6), plate (7), and neck (9) into pieces for stew or soup or bone them for ground meat. Broken line shows location of bone in chuck and foreleg.

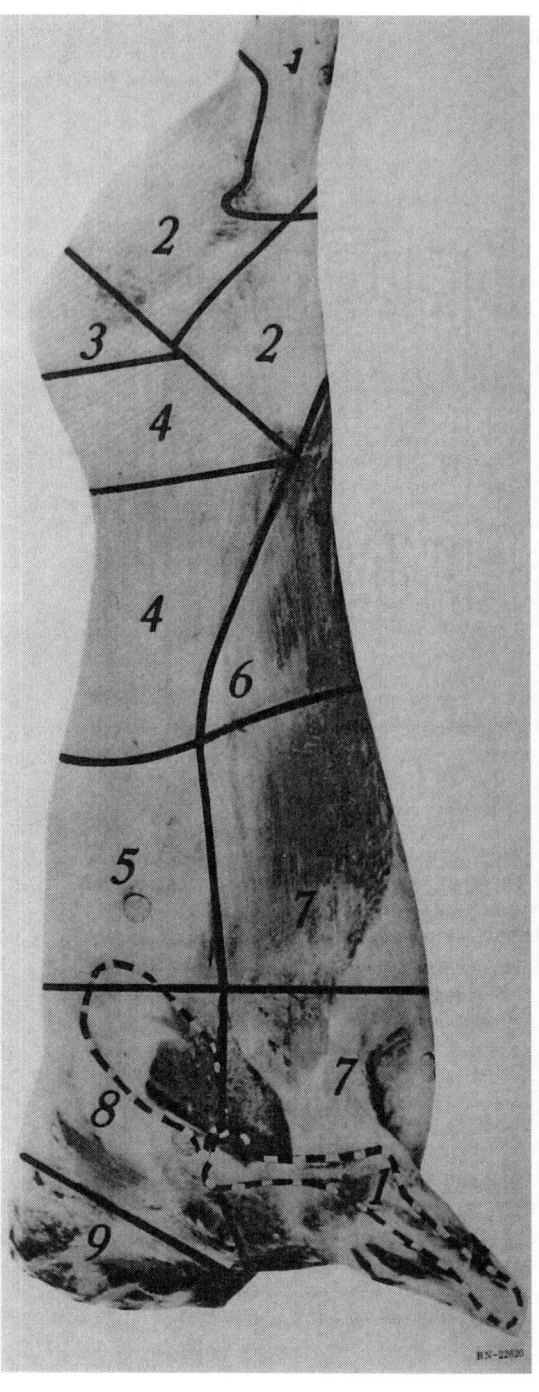

196 (4) Freezing Meat and Fish in the Home

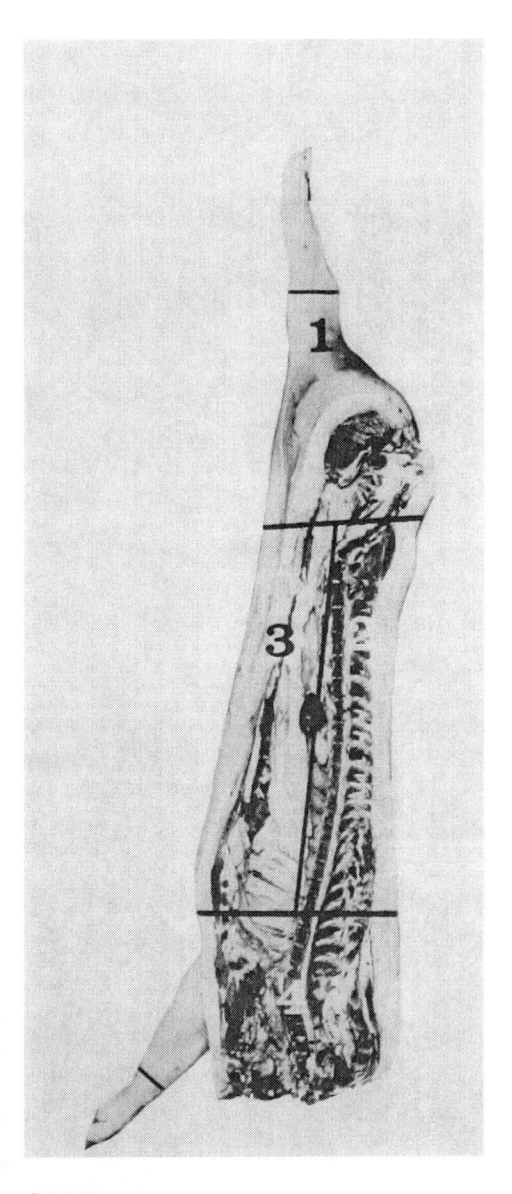

LAMB

Trim the legs (1) and shoulders (4) into uniform-sized roasts; cut rib (3) and loin (2) into chops; bone breast (5), shanks, and neck for stew or ground lamb.

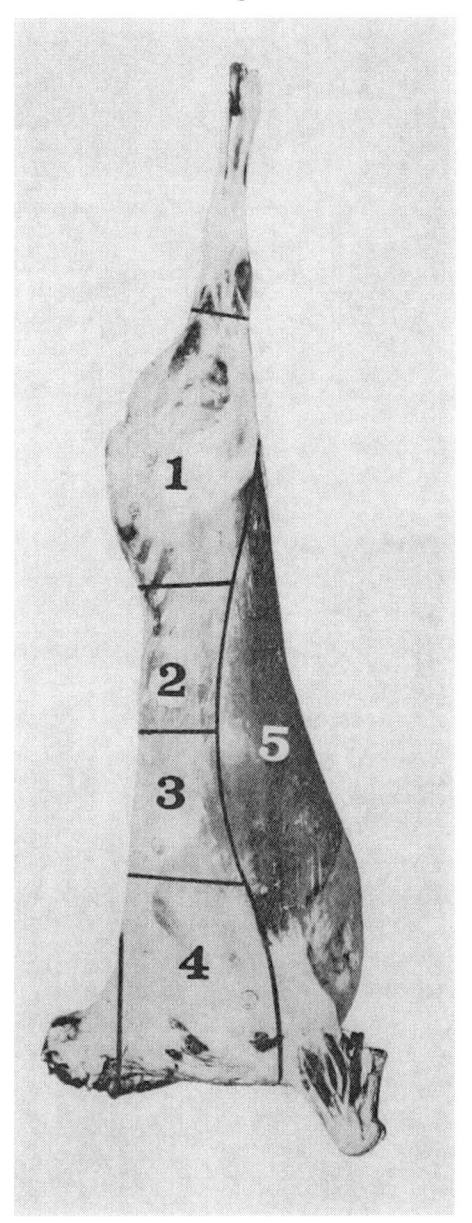

PORK

Cut or slice the ham (1), loin (2), and shoulder (4) into roasts, steaks, or chops. Trim the thin bacon strip (3) for curing, or cut into boiling pieces. Trim all meat closely, using lean for sausage and fat for lard.

Freezing Meat and Fish in the Home (5) 197

BONING THE CUTS

To save storage space, bone the cuts or trim them into smooth, compact pieces. Directions for boning and trimming are given below.

BEEF AND VEAL

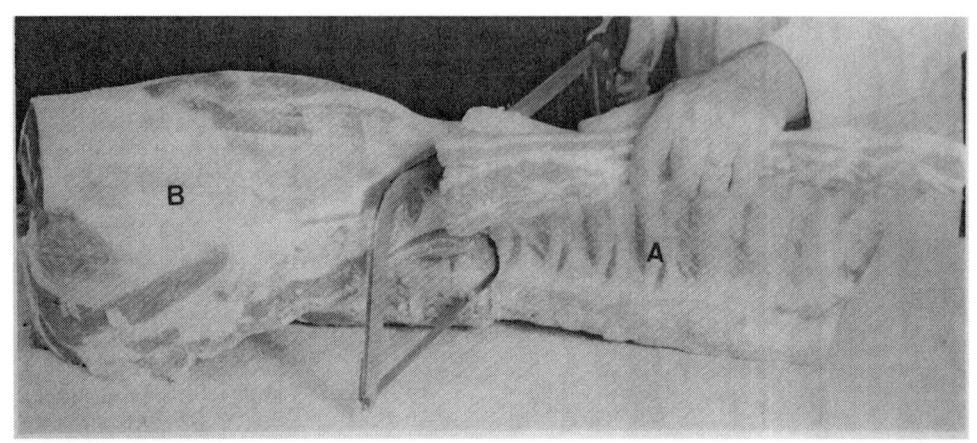

Cut the thick rib roast (A) from the chuck (B).

Remove bone from chuck for boneless roast.

198 (6) Freezing Meat and Fish in the Home

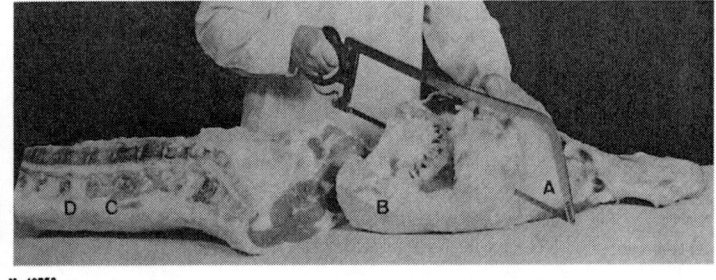

Cut round (A), rump (B), sirloin (C), and T-bone steaks (D).

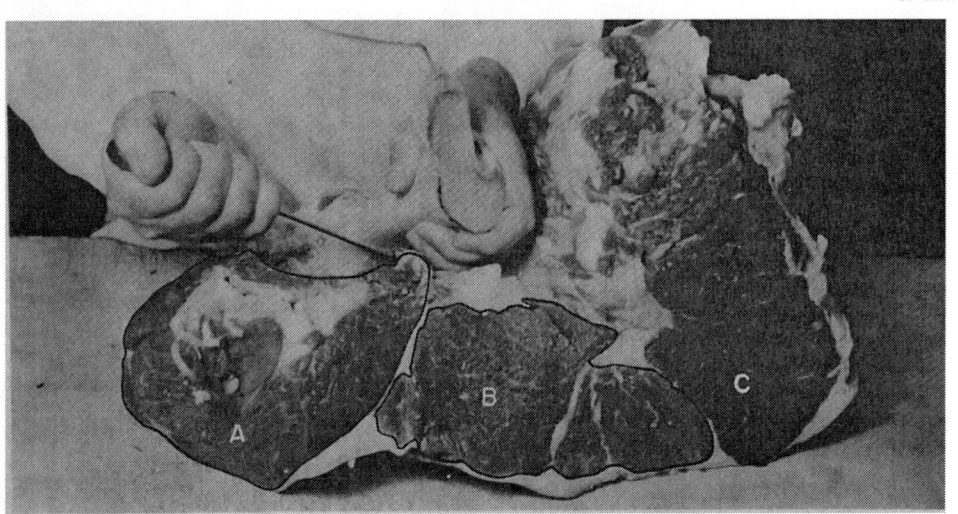

Remove bone from round and cut into tip (A), top (B), and bottom (C) round.

The result: Compact, smoothly trimmed cuts, ready for wrapping.

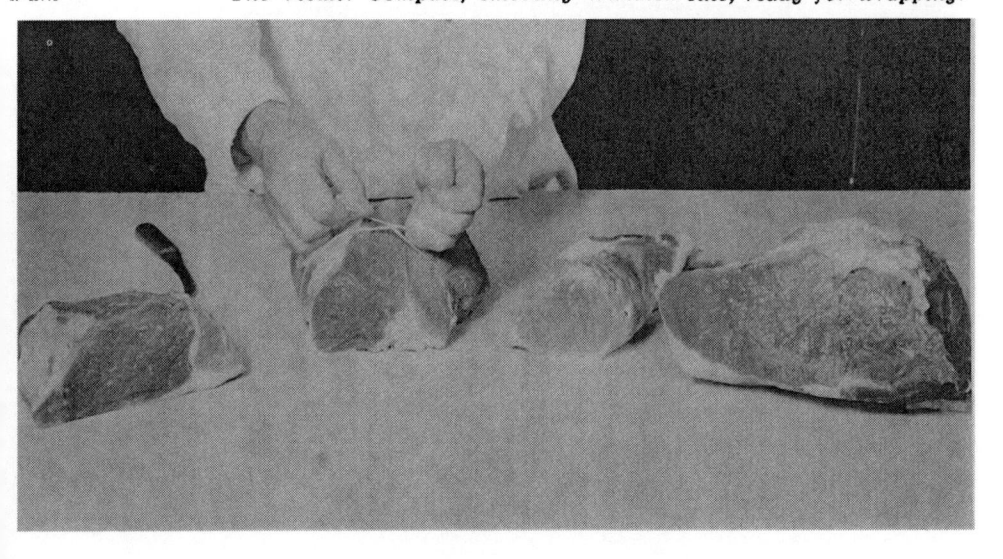

Freezing Meat and Fish in the Home (7) 199

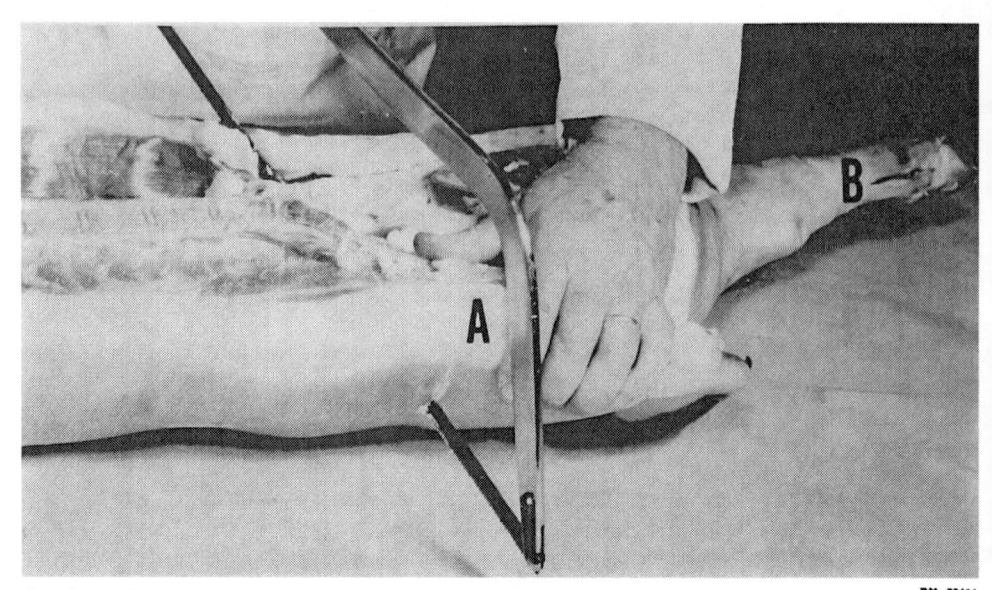

PORK

Cut ham (A) at right angles to hind shank (B).

Remove the two-rib shoulder (A).

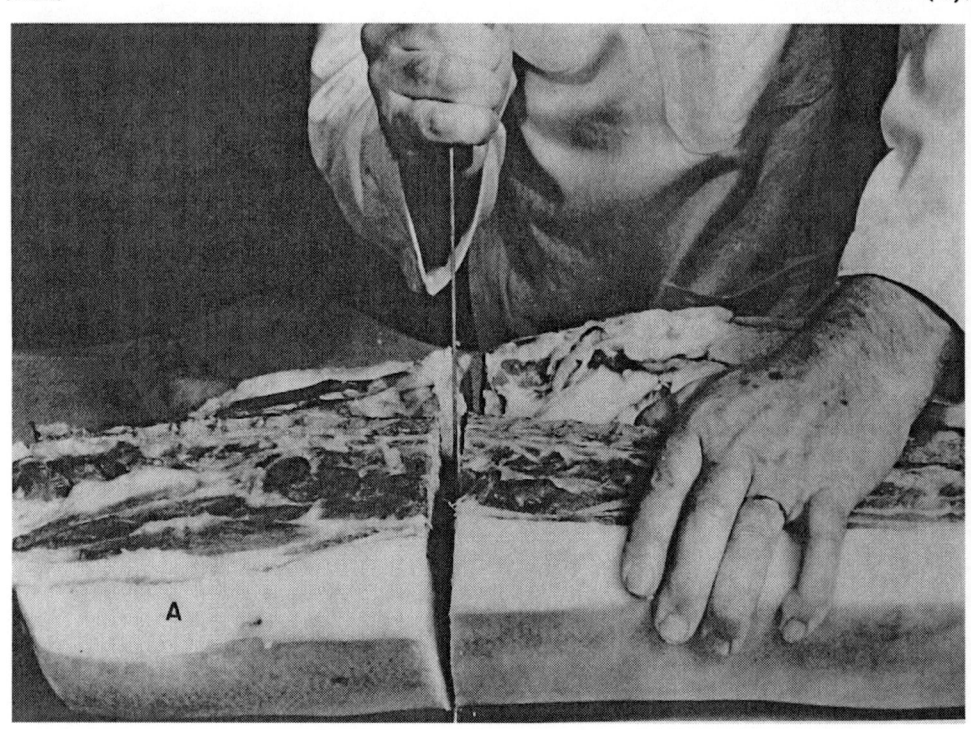

200 (8) Freezing Meat and Fish in the Home

Cut the thick loin (A) from the bacon (B).

Remove spareribs from the bacon.

Trim ham (A), bacon (B), and shoulder (C) smoothly. To save freezer space these cuts can be cured and smoked.

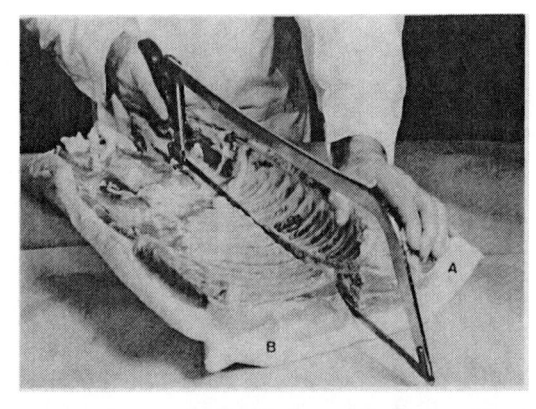

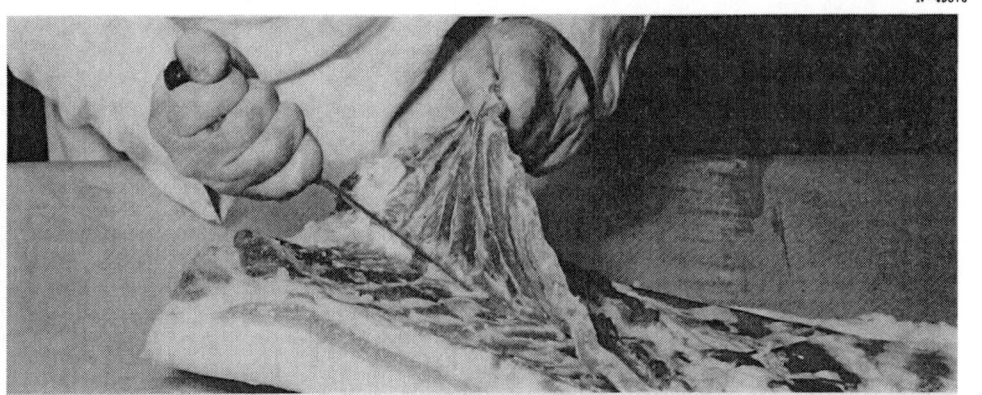

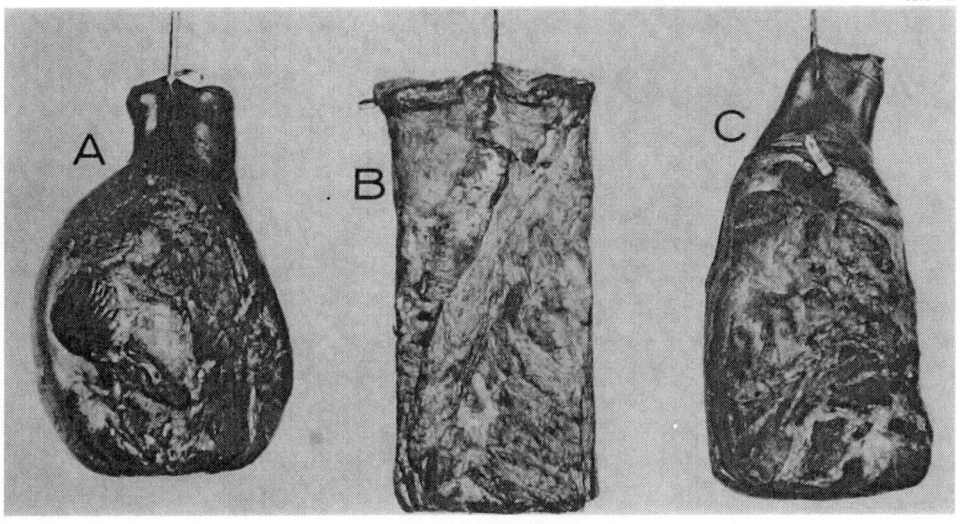

Freezing Meat and Fish in the Home (9) 201

LAMB

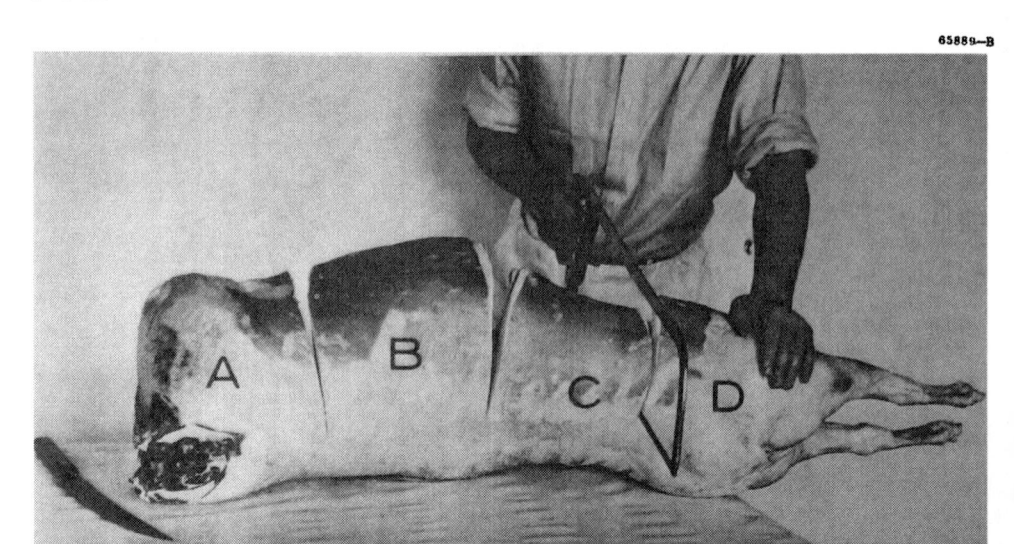

Cut to give: A five-rib shoulder (A), leaving other ribs on rack (B), loin (C), and leg (D).

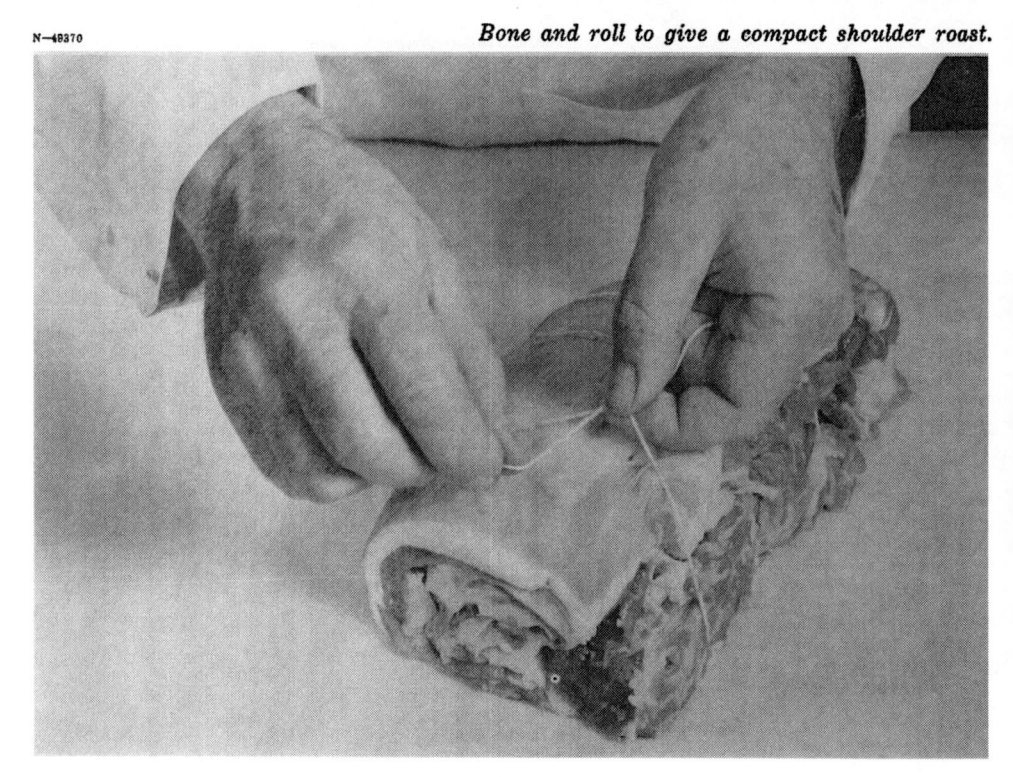

Bone and roll to give a compact shoulder roast.

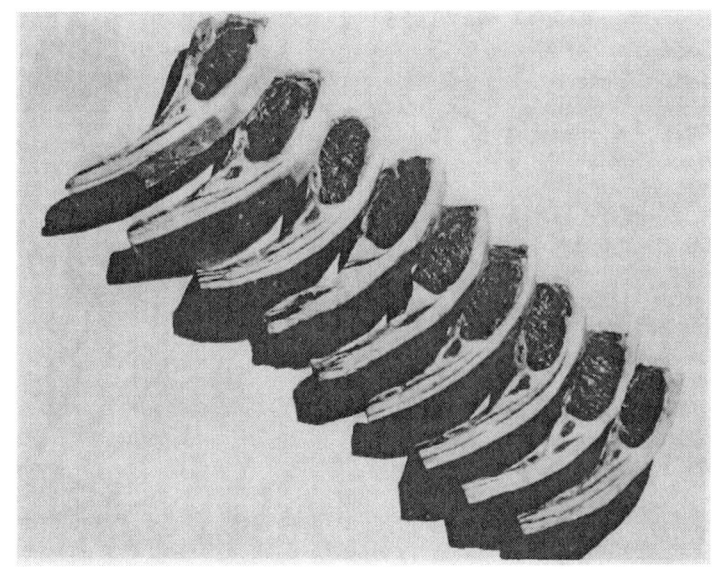

Prepare short-cut chops to save storage space.

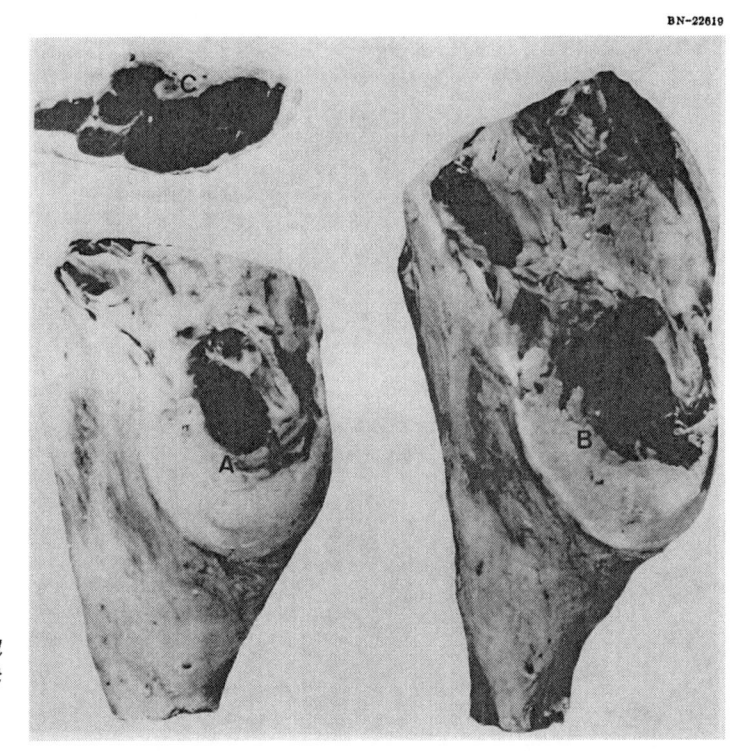

Trim leg roasts (A and B) smoothly, and cut leg chops (C).

Freezing Meat and Fish in the Home (11) 203

MEAT YIELDS

Owners of home freezers and users of freezer lockers are interested in the approximate number of pounds of various meat cuts that can be obtained from a given slaughtered meat animal or dressed carcass.

In the accompanying tables, examples are given of the approximate trimmed cut yield from slaughtered and chilled beef carcass (table 1), forequarters (table 2), hindquarters (table 3), slaughtered and chilled hog carcass (table 4), and slaughtered and chilled lamb carcass (table 5).

All trimmed meat cut yields are figured after surplus fat and bone have been removed from carcasses.

TABLE 1.—Approximate yields of trimmed beef cuts from animal having a live weight of 750 pounds and a carcass weight of 420 pounds

Trimmed cuts	Yield	Live weight	Carcass weight
	Pounds	*Percent*	*Percent*
Steaks and oven roasts	172	23	40
Pot roasts	83	11	20
Stew and ground meat	83	11	20
Total	338	45	80

TABLE 2.—Approximate yields of trimmed beef cuts from dressed forequarters weighing 218 pounds

Trimmed cuts	Yield	Weight of forequarters
	Pounds	*Percent*
Steaks and oven roasts	55	25
Pot roasts	70	32
Stew and ground meat	59	37
Total	184	84

TABLE 3.

Approximate yields of trimmed beef cuts from dressed hindquarters weighing 202 pounds

Trimmed cuts	Yield	Weight of hindquarters
	Pounds	Percent
Steaks and oven roasts	117	58
Stew, ground meat, and pot roasts	37	18
Total	154	76

TABLE 4.

Approximate trimmed pork cuts from a hog having a live weight of 225 pounds and a carcass weight of 176 pounds

Trimmed cuts	Yield	Live weight	Carcass weight
	Pounds	Percent	Percent
Fresh hams, shoulders, bacon, jowls	90	40	50
Loins, ribs, sausage	34	15	20
Total	124	55	70
Lard, rendered	12	15	27

TABLE 5.

Yields of trimmed lamb cuts from a lamb having a live weight of 85 pounds and a carcass weight of 41 pounds

Trimmed cuts	Yield	Live weight	Carcass weight
	Pounds	Percent	Percent
Legs, chops, shoulders	31	37	75
Breast and stew	7	8	15
Total	38	45	90

FISH

SELECTION

Fish may be frozen in a variety of forms or cuts as shown here:

Whole—as they come from the water.

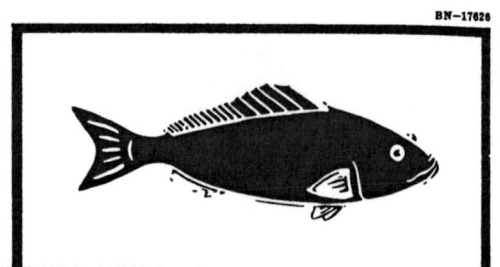

Drawn—whole fish with entrails removed.

Dressed or pan-dressed—whole fish with scales and entrails removed, usually with head, tail, and fins removed.

Steaks—cross-section slices from large dressed fish.

Fillets—sides of the fish, cut lengthwise away from the backbone.

CLEANING AND DRESSING

Usually it is best to freeze fish in package form. For home use, packaged fish has two advantages (1) it takes considerably less storage space because waste portions—head, entrails, backbone, fins, and tail—have been removed, and (2) it can be packaged in the right amount to fit family needs for one meal.

Follow these steps in cleaning and dressing fish:

1. Wash fish. Remove scales by scraping the fish gently from the tail to the head with the dull edge of a knife.
2. Remove the entrails after cutting the entire length of the belly from the vent to the head. Remove the head by cutting above the collarbone.
3. Break the backbone over the edge of the cutting board or table.
4. Remove the dorsal or large back fin by cutting the flesh along each side, and pulling the fin out. Never trim the fins off with shears or a knife because the bones at the base of the fin will be left in the fish.
5. Wash the fish thoroughly in cold running water. The fish is now dressed or pan dressed, depending on its size.

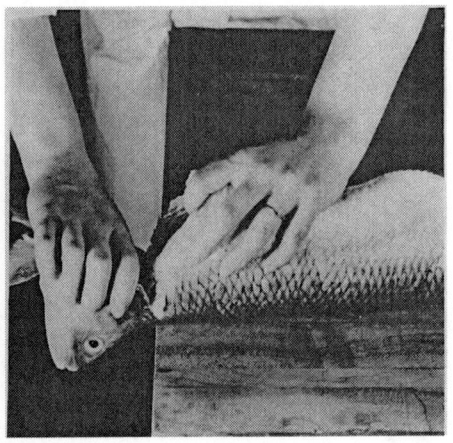

1. Scaling fish.

2. Removing head.

3. Breaking backbone.

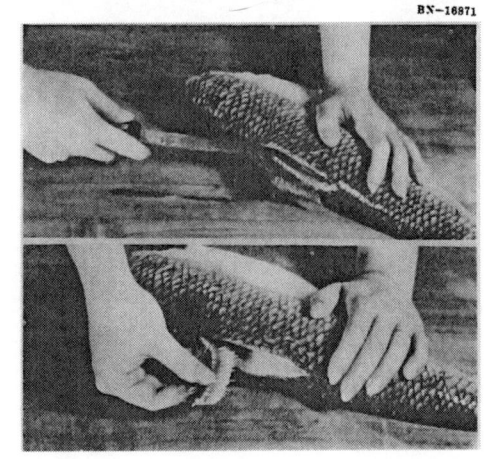

4. Removing dorsal fin.

Freezing Meat and Fish in the Home (15) 207

6. Large dressed fish may be cut crosswise into steaks for freezing, if desired. Cut steaks about ¾ of an inch thick.

7. To fillet: With a sharp knife, cut down the back of the fish from the tail to the head. Then cut down to the backbone just above the collarbone. Turn the knife flat and cut the flesh along the backbone to the tail, allowing the knife to run over the rib bones. Lift off the entire side of the fish in one piece, freeing fillet at the tail. Turn the fish over and cut fillet from the other side.

8. If you wish, you may skin the fillets. Lay the fillet flat on cutting board, skin side down. Hold the tail end with your fingers, and cut through the flesh to the skin. Flatten the knife on the skin and cut the flesh away from the skin by running the knife forward while holding the free end of the skin firmly between your fingers.

9. Commercial practice indicates that it is a good idea to give fish steaks and fillets a 30-second dip in a 5-percent salt solution before wrapping and freezing. To make a 5-percent salt solution, use ⅔ cup salt to 1 gallon of water.

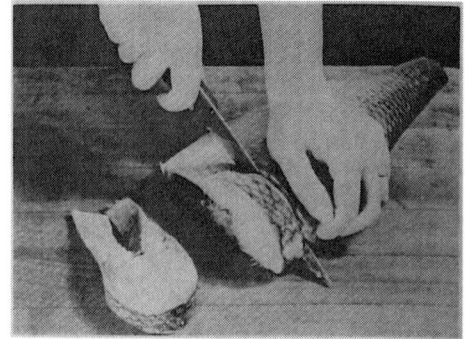

5. Cutting fish steaks.

6. Cutting from tail to head.

7. Cutting along backbone.

208 (16) Freezing Meat and Fish in the Home

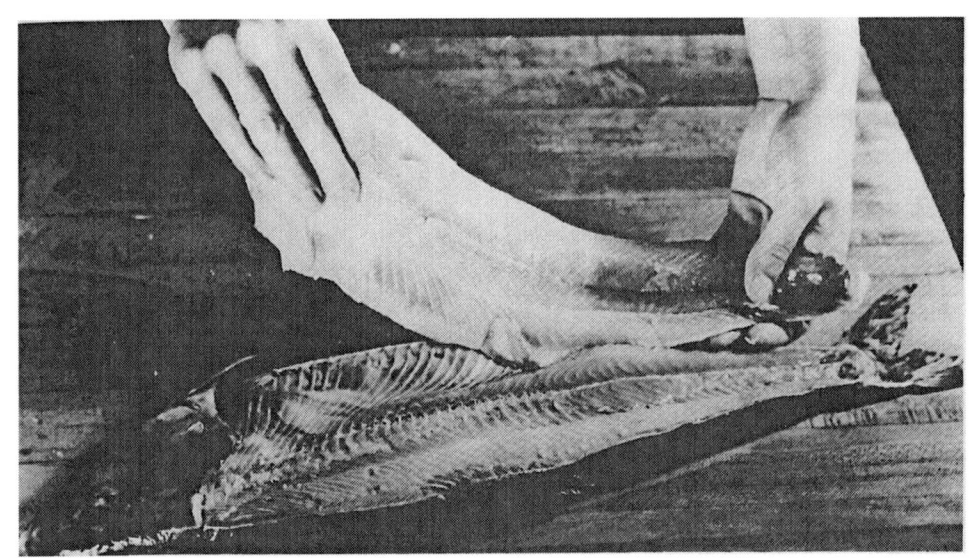

8. *Lifting out fillet.*

FISH YIELDS (APPROXIMATE)

		Edible portion (Percent)
🐟	Whole	45
🐟	Drawn	48
🐟	Dressed or pan dressed	67
🐟	Steaks	84
🐟	Fillets	100

WRAPPING, FREEZING, AND STORING

Wrap meat and fish to be frozen in moisture-vapor-resistant coverings to make the package airtight and prevent drying. Place two layers of waxed paper between individual chops, steaks, and fillets so that individual frozen pieces can be separated easily, then wrap in freezer paper.

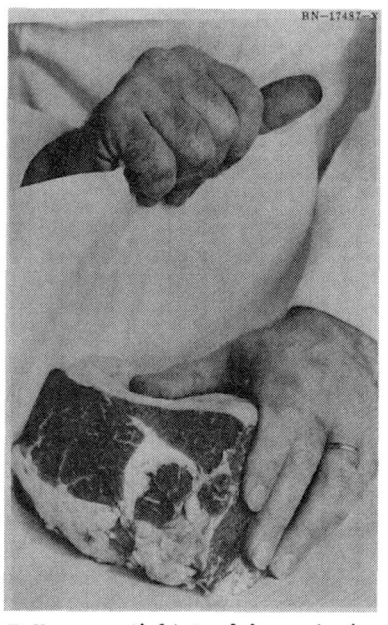

Pull paper tight to drive out air; make packages smooth to pack together snugly.

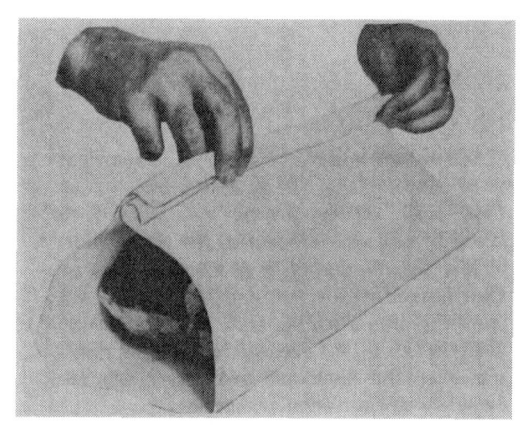

Seal seam with fold.

Fold or twist ends. Tape ends and seams.

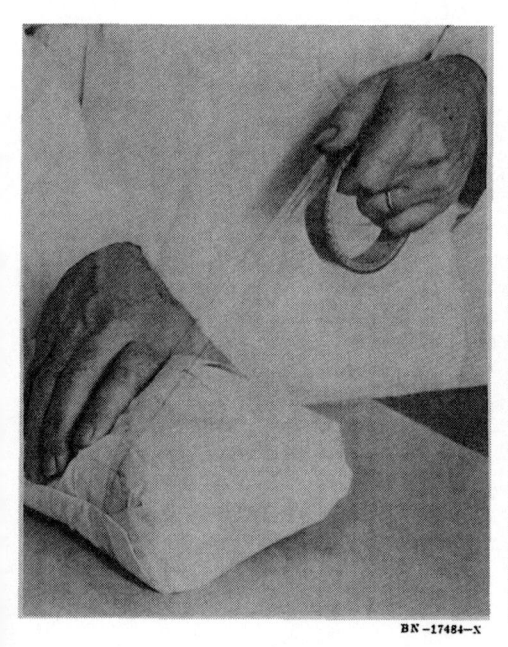

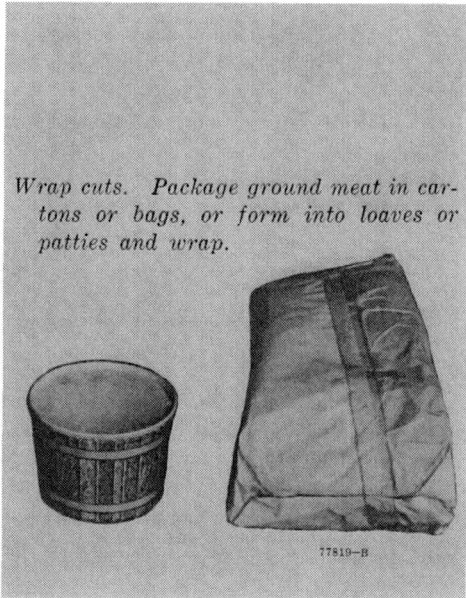

Wrap cuts. Package ground meat in cartons or bags, or form into loaves or patties and wrap.

Freezing Meat and Fish in the Home (19) 211

Spread packages to freeze. Turn control to coldest position (0° F. or lower) so that food will freeze promptly. Frost-free freezers have a built-in fan for circulation of air. If you do not have a built-in fan in your home freezer, a suggested method for speeding up freezing is to place a small electric fan in the freezer to keep air moving over the packages until they are reduced to storage temperature.

Store at 0° F. or lower.

STORAGE PERIODS

The recommended storage periods for home-frozen meats and fish held at 0° F. are given below. For best quality, use the shorter storage time.

Product	Storage period (months)	Product	Storage period (months)
Beef:		Pork, cured:[1]	
Ground meat	2 to 3	Bacon	Less than 1
Roasts	8 to 12	Ham	1 to 2
Steaks	8 to 12	Pork, fresh:	
Stew meat	2 to 3	Chops	3 to 4
Fish	6 to 9	Roasts	4 to 8
Lamb:		Sausage	1 to 2
Chops	3 to 4	Veal:	
Ground meat	2 to 3	Cutlets, chops	3 to 4
Roasts	8 to 12	Ground meat	2 to 3
Stew meat	2 to 3	Roasts	4 to 8
		Organ meats	3 to 4

[1] Frozen cured meat loses quality quickly and should be used as soon as possible.

THAWING MEATS AND FISH

Most frozen meats and fish may be cooked either with or without previous thawing. But extra cooking time must be allowed for meats not thawed first—just how much will depend on the size and shape of the cut.

Large frozen roasts may take as much as 1½ times as long to cook as unfrozen cuts of the same weight and shape. Small roasts and thin cuts, such as steaks and chops, require less time.

Frozen fish, fillets, and steaks may be cooked as if they were in the unfrozen form if additional cooking time is allowed. When fish are to be breaded and fried, or stuffed, it is more convenient to thaw them first to make handling easier.

Thaw frozen meats and fish in the refrigerator in their original wrappings.

For best quality, cook thawed meat and fish immediately.

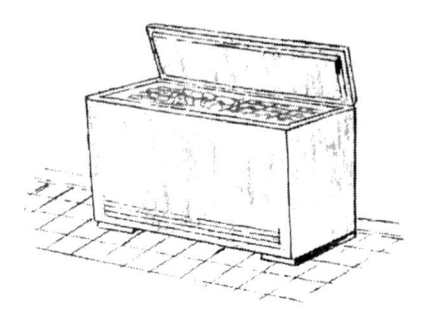

ECONOMY

For *economy*, use freezer to capacity. A well-filled freezer operates more efficiently than one that is partly filled.

QUALITY

For *quality*, keep foods moving in and out of the freezer. Put food in freezer frequently. Take food out regularly and within the recommended storage period.

CONVENIENCE

For *convenience*, you may want to keep a record of frozen foods in storage. Enter in this record the date a food went into the freezer and the date by which it should be used. When you take the food out, cross out the entry. Place the record near the freezer for easy reference.